THE

NEW

UNIVERSE

COSMOS IS CALLING

Revised 2nd Edition

ROBERT HOUSTON

Printed & Distributed by Createspace.com
Also available at Amazon.com

Second Edition; August 2017. ISBN-13: 978-0692912621

Library of Congress Cataloging-in-Publication Data

Houston, Robert, 1940-
The New Universe; Cosmos is Calling / Robert Houston
Library of Congress Catalog Number; 2017910430
BISAC: Science / Cosmology

Printed in the United States of America

This book is dedicated to all those who made a positive contribution to our scientific knowledge in order to illuminate our minds, to help elevate humanity above the darkness of ignorance and indoctrination and to help erase the barriers of discrimination, hate and bigotry.

CONTENTS

FOREWORD
TO
2nd EDITION

Lately, it seems, not a day goes by without some news of an exciting new scientific discovery in the field of physics, chemistry, biology, astronomy or medicine. With each discovery, a missing piece of the great puzzle of human knowledge falls in place to open a new window into the unknown in the scientific realm. Although these new discoveries may well be significant enough to stand up on their own merits in their own fields, they do rarely go beyond the limits of the field they belong to in order to impact other fields of science... For instance, in Particle Physics, a significant new particle may be discovered to change our understanding of the Standard Model and yet the implications of such discovery on other fields of science such as Medicine or Chemistry seem not fully warrant an investigation or understanding. We, as scientists, appear to be content with the excitement generated by our discovery, whatever that may be.

If for a moment we can assume science to be a big jigsaw puzzle where each significant new discovery represents a piece to fall in place in order to complete the big picture, the significance of the shape of one piece to the shape of another is decisive in successfully solving the puzzle to fully see the emerging picture. Similarly, in science, each new discovery regardless of the field it belongs to has an intimate and undeniable connection to everything we have so far discovered or will discover in the future in all branches of science. Yet, scientific discoveries so far in each and every branch of science seem to stand alone and advance without any in depth connection to discoveries in other fields. For example, how do our generally accepted parameters of the Standard Model impact our understanding of the processes leading to formation of a Black Hole or a Supernova? Similarly, how splicing a human gene in order to alter its structure is affected by the presumed discovery of the Higgs Boson? We all agree that there is undisputable evidence of evolution. But why there is evolution? What drives evolution to happen? What is the mechanism in cell level that allows it to happen? How our Standard Model contributes to that process we call evolution? So far however we don't appear to be interested enough to reason at that level of scientific curiosity, inquiry or understanding.

It is increasingly clear that there is an unbreakable intimate scientific connection between the smallest atomic particle we have so far discovered and the largest of celestial bodies we have so far observed in the Universe and everything in between and that the nature of the smallest dictates the nature of the largest and everything in between and vice versa. Validating and documenting that intimate connection in my opinion is the remaining big challenge for scholars to tackle as we move into the future. Accordingly, this new edition of The New Universe aspires at some level to be a small step in that direction.

The Universe is simply too big for a single mind to comprehend. I say this because I am aware of all the contributions made to science by the great minds of the past. Those were minds that had awesome power to illuminate our way through the ages in order to enrich our scientific knowledge. I am forever indebted to them for giving me the knowledge so I could be educated without being indoctrinated. As a result, this book did not attempt to search for an explanation for every observation or event in the Universe. Instead, its objective was to form a framework from which further scientific inquiry can be launched in order illuminate the way for more significant discoveries and help answer some of the most persistent questions about who we are and why we are here. It is natural to expect that there will be differences of opinions about a myriad of subjects mentioned in this book. I welcome dissent where and when plausible as long as it is part of our efforts to advance our understanding of the Universe, our place within and our destiny. After all, we, as humanity, are a piece in the great jigsaw puzzle we call Universe and have an obligation, in my opinion, to help elevate humanity above darkness, to protect our environment and the species within as well as help improve our lives. Without advancing our scientific knowledge, this will be a difficult task

This second edition expands on the ideas presented in the first edition and includes new concepts both at the atomic and cosmic scale. A desire to search for knowledge is behind the every word written in this work. If, by any chance, this book inspires you to see the Universe in a new light, I would consider that its mission has been accomplished.

With that in mind, let the journey begin.

FOREWORD
TO
1ˢᵗ EDITION

If I have to describe Dr. Wernher von Braun in a single sentence, I guess it will be historically correct to name him as the first person who freed humanity from the entrapment of Earth's gravity. The launch vehicles he had designed and developed, allowed, for the first time, interplanetary flights such as the Voyager Missions, Moon Landings and the placement of Hubble Telescope in an orbit above the undesirable filtering effects of our atmosphere. Such accomplishments resulted in an unprecedented expansion of our observations about the planet we live on, our Solar System and the Universe. Our efforts to interpret these new observations opened the door for a large number of new theories and speculations about everything around us including the creation of the Universe, in part, to satisfy our need and desire to assign a beginning and an end to every event we witness.

However, there appears to be an aspect of the Universe, which would not reveal its secrets, yet dictates that we proceed in an orderly fashion, sort of, one step at a time, to confirm our beliefs. If this perception is correct, then it would indeed be logical to pursue a higher level of understanding of Galaxies, including their formation, evolution and disintegration in order to journey into the farthest frontiers of the Universe. This book, then, is a small step in that direction.

All books have a reason to exist. The reason behind this book is an event that took place in the summer of 1994. In two separate letters to the

American Geophysical Union and the Society of Exploration Geophysicists, I first raised the possibility of faster rotation of the Earth's inner core than its outer shell. At the time, this idea was beyond the generally accepted standards of our scientific world. In 1997, however, the scientists documented the first evidence to support the validity of this idea. Similarly, there are ideas presented in this book that are now beyond our current scientific realm. Hopefully, these ideas will lead to some significant discoveries that will enhance our understanding of the Universe and will enrich your enthusiasm about our scientific knowledge.

I take pleasure in welcoming you aboard.

ACKNOWLEDGEMENTS

This book could not have been possible without the work of all scientists before me since the dawn of humanity. I am greatly indebted to all of them for helping me to understand.

SPECIAL THANKS

Special thanks to NASA and Wiki Commons for offering their images under public domain rules, which made it possible those images to be used in this book. I also wish to thank my daughter Stephanie for her assistance in making the digital illustrations possible. Without her help, this book could not have been complete.

If you wish to create
an atom,
you must first create
a Universe.

CHAPTER 1

SPECTRAL THEORY OF ATOMS

1.1 Introduction

Ancient civilizations had long wondered about the nature of matter and had considered the subject by offering concepts and explanations heavily influenced by the philosophical and spiritual beliefs of their times. In the 5th Century B.C., a Greek philosopher named Democritus, for the first time, asserted that the matter was made of very small indivisible particles called atoms despite the fact that these atoms were too small to be seen by naked eye. Although hotly contested at the time, his concept eventually gained acceptance and remained as the prevailing theory of understanding the nature of matter until the 19th Century.

In the early years of the 19th Century, an Englishman and a scientist, John Dalton (1766-1844) proposed his own atomic theory, in which, he suggested that each element known at the time was composed of atoms of a single and unique type and these atoms were fundamental, indivisible and indestructible even though they could combine to form other chemical compounds in reversible chemical reactions. His proposals were helpful in understanding the first law of conservation of mass as defined in 1789 by Antoine-Laurent Lavoisier (1743-1794), a French nobleman known today as the father of modern Chemistry, as well as the law of definite proportions formulated in 1799 by Joseph Louis Proust (1754-1826), a French chemist, which stated that if a compound is broken down into its constituent elements, the masses of constituents will always have the same proportions.

In 1897, however, a new discovery brought an end to the fundamentally indivisible nature of atoms. While experimenting with Crooke's tube, which is a sealed glass vacuum container with an anode and a cathode, Sir Joseph John Thomson (1856-1940), a British physicist and a Nobel laureate, observed a glowing beam of light between electrodes when a high voltage electric potential is applied. He called this glowing beam of light a Cathode Ray. Through experimentation with electric and magnetic fields, which affected the path of the Cathode Ray, he concluded that the Cathode Rays were negatively charged particles, emitted by the very atoms of the cathode. These negatively charged particles will later be called electrons. Presence of electrons meant that atoms were divisible and electrons were part of the atomic structure. These observations eventually led Sir Thomson to conclude that electrons were in fact floating in a cloud of positive electricity, which in turn made the atoms neutral in their behavior. Incidentally, this atomic model was named as the Plum Pudding Model by other scientists of the era despite opposition by Sir Thomson.

In 1909, Ernest Rutherford (1871-1937) working with Ernest Marsden (1889-1970) and Johannes (Hans) Wilhelm Geiger (1882-1945) conducted an experiment during which positively charged alpha particles were shot at a gold foil. Initially referred to as Rutherford Experiment, this observation which later became to be known as Geiger-Marsden Experiment, indicated that most of the atomic mass was concentrated in the center of an atom. This discovery later led Danish physicist Niels Henrik David Bohr (1885-1962) to conclude that the heavy nucleus of an atom is orbited by point like electrons in a manner similar to motions of planets around our Sun.

In 1932, Sir James Chadwick (1891-1974) exposed hydrogen and nitrogen to beryllium radiation and by measuring the energies of recoiling charged particles; he deduced that the radiation was actually composed of electrically neutral particles with a mass similar to that of a proton. He called these particles neutrons. Nevertheless, discoveries of subatomic particles did not stop there. Scientists, using particle accelerators, continued to discover subatomic particles in significant numbers and their work eventually led to a new and generally accepted atomic model, coined as The Standard Model.

1.2 The Standard Model

According to the standard model, all subatomic particles are generally members of two basic groups. Heavy particles, sometimes referred as Hadrons, are said to include neutrons and protons, which in turn, are made up of quarks. The light particles, called Leptons, are said to include electrons, muons and tau particles. Quarks and Leptons are considered to be a subset of a bigger group called Fermions, named after Italian physicist Enrico Fermi (1901-1954), who is known to have made great contributions to Particle Physics. In addition, there is said to be a third group of particles, called Bosons which are said to be force carriers. This group is also proposed to include gluons, the carriers of the strong nuclear force, which help quarks to combine and stay together.

 Among of those mentioned so far, quarks, in turn, are said to have 6 different types; conveniently called flavors, which are, up, down, strange, charm, top and bottom quarks. The last two are also known as truth and beauty respectively. In addition, each of these quarks are said to have their anti-particles, which in turn, are called antiup, antidown, antistrange, anticharm, antitop and antibottom. According to standard model, quarks also carry an electric charge. For instance an up quark has a positive electric charge of 2/3 of an elementary charge whereas a down quark has a negative 1/3 charge. Now, according to the standard model, two up quarks and a down quark combine with the help of gluons to form a proton, which yields to a total electric charge of +1 for the proton since $(+2/3)+(+2/3)+(-1/3)=1$. Similarly, an up quark and two down quarks combine to form a neutron whose electric charge equals to zero since $(-1/3)+(-1/3)+(+2/3)=0$. While three quarks are said to unite to form a Baryon, which is another name for protons and neutrons, a quark and an antiquark unite to form a Meson, which is considered to be a composite particle and a boson and is said to be involved in the interactions which holds the nucleus together.

As for the Leptons, there are also said to be 6 types, which are, electron, electron neutrino, muon, muon neutrino, tau particle and tau neutrino. Muons and tau particles are said to be like electrons but have more mass than electrons whereas neutrinos are said to be particles with a very small mass at the atomic scale and no charge, but they are said to carry energy. Neutrinos are said to be observed during a process called beta decay. Leptons are also said to be known to have their own antiparticles.

According to Standard Model, there are four known types of Bosons, which are said to be force carriers. They are, namely, Photons which carries the electromagnetic force, Gluons, which carries the strong nuclear force, which in turn, keeps quarks bonded together, W and Z Bosons as weak nuclear force carriers that are said to be involved in chemical reactions and finally the gravitons, which carries the force of gravity. The gravitons are not currently included in the standard model since their existence is theoretical and unconfirmed. Unlike quarks and leptons, bosons do not have any antiparticles. Photons and gluons are known to have no mass while weak force bosons have small amount of mass. While fermions are known to have half spins, bosons are said to have full spins expressed in integers.

At the time of this writing, there are said to be six types of quarks, six types of leptons and four types of bosons that make up the standard model. However, new subatomic particles are discovered every day and the list goes on. For instance, at the time of this writing, there are about 140 mesons whose existences are confirmed, and yet, more mesons are expected to be discovered and added to the list. A particular interest is a force carrier called Higgs Boson which is also referred by some as the God Particle. The Higgs Boson is suspected to interact with particles without any mass to cause them to become particles with mass. At the time of this writing, there are claims that this elusive particle has finally been confirmed to exist by scientists at the Large Hadron Collider built and operated by CERN which stands for European Organization For Nuclear Research.

Nevertheless, there are questions about the standard model that need to be investigated. For instance, quarks can only exist within a bag such as a proton or a neutron and have not been observed in isolation in nature. Accordingly, how is it possible for them to gather in various combinations to form composite particles such as baryons and mesons if they can not stand alone? There are also particles whose half lives are in nanoseconds. These particles can not exist isolated in nature long enough to participate in the formation of any atom. If that is the case, how is it possible for these particles to be a part of any subatomic structure? Furthermore, how does an atom form in the nature or in the Universe? What kind of mechanism should exist in the universe to fabricate an atom that is so complex and what should be the specifications and properties of such a mechanism?

1.3 Blacksmith Experiment

It is clear that the information so far we know about the standard model makes it nearly impossible to be used as a basis for the investigations we would like to conduct within the framework of this book. For this reason, we will go back to an experiment that long puzzled the classical physicists. In this experiment, a blacksmith heats up a metal rod until it glows with radiant energy, first with a red tint, later, at higher temperatures, with a yellow tint as shown in Figs: 1-1a & 1-1b. It is interesting to notice that the most heated part of the metal glows yellowish white and the relatively cooler portion glows with a reddish tint. If we compare this color display of the metal rod with the spectrum of visible light, we immediately notice that yellow light has a higher frequency than the red light. This means, beyond any reasonable doubt that higher frequencies are associated with higher temperatures. Furthermore, as the temperature drops, the frequency of the glowing light also drops and this manifests itself in reddish tint of the radiant energy. Finally, when the temperature drops further, the radiant energy disappears and the metal rod no longer glows. If we repeat this experiment over and over again, the same phenomenon repeats itself without any change in the physical and chemical composition of the metal rod. As a result, we can deduce significant conclusions from this experiment. First of all, the atoms subjected to heating are not chemically altered. The radiant glow is exactly same in its appearance and color in all directions meaning it is not affected by the direction of observation and only thing that seems to affect the color of the radiant energy is the temperature. The question then becomes, what in the atomic structure of the heated metal rod is capable of radiating the frequencies of the spectrum of the visible light? Is it possible to explain this radiant energy using principles of the quantum theory?

It is generally accepted as a fact that according to the quantum theory, energy is emitted in discrete amounts called quanta when electrons jump between orbits of different energy levels surrounding the nucleus. Since this process is expected to be purely random in nature, resulting frequency-amplitude signature of the radiant energy should be different than the consistent frequency-amplitude spectrum of the visible light. And yet, there seems to be a proportional relation between frequency spectrum of the radiant energy and temperature In addition, the Ultraviolet Catastrophe suggested by the Raleigh's law is completely inconsistent with the purely electromagnetic nature of the visible light radiated in our experiment since electromagnetic spectrum is a continuum between gamma rays and the radio waves. Furthermore, there is so far no evidence of existence of any blackbody particle that can absorb and re-emit all radiation as proposed and used by Gustav Kirchhoff (1824-1887) in his explanation of the radiant energy observed when a piece of metal is heated. Even more questionable is the assumption made by Max Planck (1858-1947) that the radiant energy was emitted by tiny resonators within the atomic structure since there has been no evidence so far of such a particle within the atomic structure as presented by the standard model. In addition, Standard Model does not address the wave and particle nature of visible light and the efforts to discover a particle that generates frequencies within the atomic structures have so far failed to yield a conclusive result.

Based on these considerations, it becomes necessary to approach to the understanding of the radiant energy illustrated in our experiment from a completely different point of view.

Using the principles of reverse engineering, we will try to establish what an atom should be like and compare its properties with the results of well-established physical experiments instead of working with the Standard Model. Admittedly, this is no simple task because we must always bear in mind all fundamental laws of Physics with which we can't be in contradiction of any kind. However, we will select the most fundamental laws of Physics to make our comparisons, correctness of which we are fairly certain.

For our specific purposes, we will require an atom with great flexibility of size and shape in response to the changing conditions of its physical environment. It should have a simple structure that can be easily formed in nature without having to require elaborate undertakings. But most importantly, it should conform to the particle and wave nature of light.

The most suitable shape for any atom is a sphere because a sphere has symmetry on its infinite number of axes and yet can be formed in nature with great ease since Cosmos is full of spherical bodies formed entirely by the forces beyond our control. (Remember the glowing iron rod? The color of glow was same in all directions) Its spherical shape should also be very flexible, very much like soap bubble floating in the air. Like a soap bubble, our atom has no nucleus since all of its contents are in its shell. The shell itself is composed of positive electricity inside, negative electricity outside and the atomic mass in between the two. Since negative electric charge is outside the shell, all chemical reactions can be performed as usual and the reactions with positive electric charges within, namely nuclear reactions, will require the destruction of its shell before they can happen. If we shoot a charged particle traveling at a very high speed at our atom and hit it, it will shutter and its fragments will scatter like those of a glass bottle hit by a bullet in a spaghetti Western. (We will revisit this assumption a little later in the book)

We all know that likes repel so the positive electric charges within our atom are trying to expand its shell by pushing it outward with a force inversely proportional with its expansion. If there is no force to stop this expansion, the shell of our atom will expand forever, albeit at an increasingly slower rate with corresponding lower frequency and amplitude. However, there is another force within our atom and that is the force of microgravity between each and every point of its mass of the spherical shell and this force is trying to contract the shell towards its center. Fig: 1-2. As a result, the shell of our atom expands and contracts, creating a volumetric oscillation, sort of like a pendulum, only in three dimensions. Fig: 1-3a and Fig: 1-3b. These volumetric oscillations of the shell of our atom move the negative electric charges on its surface back and forth in all directions to create a three dimensional electromagnetic radiation field which allows our atom to behave, not only like a particle, but also like a wave while mass of its shell generates a three dimensional gravity wave field. This wave and particle property of our atom is unchanged even when our atom is heated up to much higher temperatures to act like a Photon which allows us to propose that light is made up of atoms at higher temperatures. (We will further discuss this property of our atoms later in this book) Furthermore, the negative charges outside the shell of our atom cling to its surface and are held in place by the attraction of the positive charges within. As a result our atom does not require particles or force carriers to keep it together.

Atoms of each element have a special ratio that defines its physical and chemical properties. This is the ratio of its electric charge to its mass. I call this E/M ratio, the ratio of energy over mass. Here energy refers to the electric charges of our atom and the mass refers to the mass of its shell. We will come back to this ratio a little further down this book since it is such an important parameter of our atomic model.

Nevertheless, one important question remains; how can soft atoms form a piece of hard rock? When we squeeze a rock, we are actually applying a force against the positive charges within the atom, not on its shell. The shell itself can't withstand the force we are applying, but the electric charge that is trying to expand can since it is very strong at the atomic scale.

1.4 Interactive Atoms

What we have described above is a dynamic atom as opposed to the static nature of the Standard Model. As we will see later in the following chapters, the universe and life can't exist without this fundamental dynamic property of our atom.

Now, let us consider the case of two similar atoms placed side by side. The negative charges outside of their shells will repel each other, so our two atoms are trying to distance themselves from each other. On the other hand, force of microgravity, this time between each and every points of the atomic masses of the adjacent atomic shells are pulling these two atoms closer. Fig: 1-4. The comparative strengths of these atomic forces allow three possible outcomes of the inter-atomic balance. If the force of electric repulsion is greater than the attractive force of microgravity, then these two atoms belong to a gaseous element, such as Helium. On the other hand, if the attractive force of microgravity is stronger than the force of electric repulsion, these two atoms belong to a solid element, such as Iron. A near balance between these forces results in a liquid element, such as Mercury. This explains why there is only one liquid element, namely Mercury.

In considering these two similar atoms so far, we have assumed that these two atoms have similar physical properties, such as their geometry, size of their electric charges, mass of their shells, frequency and amplitude of their volumetric oscillations and therefore their electromagnetic fields in three dimensions. Since heat is electromagnetic radiation, we can say that these two atoms are at the same temperature level. In case of two similar atoms with different temperature levels, the frequency and amplitude of their volumetric oscillations are different meaning they have two different atomic spectra. Fig: 1-5. The atom with the higher temperature has a relatively higher frequency and amplitude, therefore a more powerful spectra and electromagnetic radiation, which overpowers electromagnetic radiation and atomic spectra of the cooler atom and in doing so it alters and elevates the frequency and amplitude of the cooler atom, raising its temperature. This, in a nut shell, is heat transfer. It also implies that there is only one kind of heat transfer, electromagnetic radiation, instead of three, namely, conduction, convection and radiation. This type of interaction is also behind the phenomenon called "Northern Lights" or "Aurora Borealis". When sun's radiation, also called Solar Winds by some scientists, increases its intensity, it excites the atoms and molecules in the upper atmosphere, increasing their spectra into the range of visible light spectrum, making their collective motion visible to the naked eye. Fig: 1-1d and Fig: 1-1e.

When the temperature of two adjacent atoms are elevated, increasing their frequency and amplitude, they now occupy a greater space in universe when they are at their maximum volume in order to provide enough space for their volumetric oscillations to take place freely. This, in turn, means that the distance between their volumetric centers are increased. This is called expansion. Fig: 1-6. As a result, when we heat a metal bar, it expands. Similarly, when we heat a gaseous element, its atoms also expand and try to occupy a greater volume. If this gaseous element is within a confined space, this force of expansion results in increased pressure and temperature because of decreased free distance between their volumetric centers and increased repulsive force between the atoms. This is possible because energy decreases by the inverse square of distance. (Reasons behind increased temperature will be dealt shortly)

Let us for a moment talk about a large number similar atoms belonging to a gaseous element. We now know that their repulsive forces are greater than their attractive forces, so the each atom of this gaseous element is trying to distance itself from all others. (The force exerted on one single atom is mainly dominated by the nearest atom in its vicinity although all atoms nearby do contribute to overall force field affecting its direction of movement) As a result, these two atoms in the nearest vicinity of each other will end up moving in opposite directions until each gets close to another atom which in turn will affect their direction of movement, forcing both to move in yet another direction as dictated by repulsive force exerted by the atoms on their path. This is "Brownian Movement", also called "Random Walk". "Diffusion" is not possible without "Random Walk.

We are now ready to discuss temperature variations on atoms. Reducing temperature will result in reduced levels of amplitude and frequency of volumetric oscillations and will result in reduced repulsion forces, allowing formation of liquids and solids from gases. The opposite of this process is also true, which means that, increasing temperature will result in formation of gases from liquids and solids.

If we can strip negative electric charges from the surface of our atom, it becomes a positively charged particle. Accordingly, our Helium atom becomes an Alpha Particle when it is stripped off its negative charge.

1.5 Spectral Theory at Extreme Temperatures

If we continue to heat our atom, its frequency and amplitude continues to increase. At very high temperatures our atom becomes a source of "Gamma Rays" and at lower temperatures it is a source of "Radio Waves".

If we subject our atom to extremely high temperatures of million degrees of centigrade or more, something unexpected happens. As the temperature increases, the E/M ratio of our atom begins to change. As the temperature increases our atom has more electric charge and less mass. In other words, the mass of our atom begins to dissolve into its components of negative and positive electricity. We do not notice this change at room temperatures since it only happens at extreme

24

temperatures of million and billion degrees. When our atom is more energy and less mass, it transforms into a state of Plasma. When temperature is further increased, our atom becomes all electric charge (energy) and no mass. At this extreme state, our atom is now a Photon. Nebulae in interstellar space, in the opinion of this author, are examples of plasma state as opposed to its more popular current definition of being a cloud of gas and dust. We will revisit this subject further in the following chapters.

At the opposite end of temperature spectrum, at extremely low temperatures, possibly millions of degrees below zero, our atom is all mass and no energy. Its positive and negative electric charges have integrated to become additional mass therefore increasing its mass to its maximum quantity. Since there is no electric charge trying to expand its shell, our atom has no frequency and amplitude and no electromagnetic radiation in a three dimensional field. Fig: 1-7. It has collapsed on itself leaving our atom occupying its smallest possible volume. At these extremely low temperatures, our atom is Dark Matter. A large percent of matter in the universe is said to be Dark matter by astronomers. We will revisit this subject when we discuss the age of Universe further down in this book.

Black Holes are examples of gatherings of atoms at very low temperatures and this explains why Black Holes have such an immense gravity signatures. Since there is no repulsive force of the negative electric charge, atoms at these extremely low temperatures have no volumetric oscillations and therefore no electromagnetic radiation. As a result, Black Holes can't be directly detected by optical or radio telescopes at great distances. However, existence of Black Holes in the Universe has been indirectly confirmed by some astronomers by means of their gravity pull on their surroundings.

1.6 Crystallization

So far we have only talked about atoms of single elements. The situation is a little different when we talk about the interaction of two different atoms of two different elements. As we have said before, at room temperature, each atom, as a function of its geometry, energy, mass and spectrum, occupies a unique volume of space. In a compound such as a mineral, where there are two or more different kinds of atoms, the space between different atoms are dictated by their comparative attractive and repulsive forces. For instance, attractive and repulsive forces between two sulfur atoms is different than, say, attractive and repulsive forces between two carbon atoms. As a result, the distance between a sulfur atom and a carbon atom will be different than the distance between two carbon atoms as well as the distance between two sulfur atoms. This difference in relative spacing of different atoms results in different crystalline forms and structures. If more elements exist within a given mineral, more complex its crystalline form and structure become. This, in general, dictates how many different dimensions exist within a crystalline form. Fig: 1-8.

1.7 Temperature Gradient of an Atom

It is clear from our considerations so far that our atom has electromagnetic radiation and that translates into heat. It also means that as we walk away from our atom, we feel less of its heat since electromagnetic radiation decreases with the inverse square of the distance between the atom and us. If we plot this temperature variation along anyone of our atom's axes, we get its temperature gradient as shown in Fig: 1-9. This is important because it implies that as we get closer to our atom, we feel more of its heat. The opposite is also true since we walk away from it, less heat we feel. This is a significant observation since it implies that if we push two atoms together, both of them will have higher temperatures. So, if we compress a gas in a container, the temperature of the gas rises. This has major implications in cosmology, especially in creation of stars and galaxies as well as in our considerations over cosmic cycle. We will come back to this subject again to discuss it in detail.

1.8 Magnetism

So far we have presented an atom, perfectly spherical in shape, whose contents are uniformly distributed in its shell. This is not always the case, however. Under certain conditions, electric charges of our atom may not be distributed evenly within and outside of its shell. When this is the case, something unusual happens. The side of our atom that has more electric charges ends up having a greater repulsive force since presence of additional electric charges now overpower the gravity pull of its mass while the side with lesser amount of electric charges develops a greater attractive force or gravity pull since its electric charges are now overpowered by the gravity pull of its mass. As a result, our atom now has polarity and spins around its vertical axis when it is subjected to an electromagnetic field to align itself with the polarity of the electromagnetic field it is in. Furthermore, its attractive and repulsive forces are not uniform in all directions which mean that it either attracts or repels certain objects in its near vicinity with different levels of force. We are not going to discuss magnetism in this book in detail since it is presented in a wide variety of science books. Our purpose here is to understand how an atom can have magnetic properties.

Magnetite, an ore of iron, has natural magnetic properties. We can also create magnetic properties by inserting an iron rod into a coil. When we run an electric current through the coil, electromagnetic field created by the coil disturbs the uniform distribution of electric charges of the atoms of the iron rod which results in its polarization. This in turn affects the balance of the attractive and repulsive forces of the atoms of the iron rod repelling like electric charges and attracting opposite electric charges with a greater force depending on its direction of polarity.

Cross section of a magnetized atom is shown in Fig: 1-10. From this figure we can clearly see that magnetic properties of our atom are the result of uneven distribution of its electric charges whereas distribution of its mass remains mostly unchanged. As a result, it is possible to assert that magnetism is nothing other than a manifestation of a lack of balance of power between gravity pull of mass and repulsive force of negative electric charges, magnified significantly. This strongly suggests that magnetic pull is micro gravity under special conditions.

From these conclusions, we can safely assume that our atom also has a gravity gradient as shown in Fig: 1-11 when its attractive forces are greater than its repulsive forces.

1.9 Unified Field Revisited

If magnetism is micro gravity which is an attribute of mass when an imbalance of electric charges exists at the atomic level, it is also a property of energy since energy and mass are two phases of matter and they are convertible between the two as expressed by Albert Einstein (1879-1955) in his famous formula of $E=mc^2$. Here c^2 is nothing more than a conversion ratio between the two. In other words, both magnetism and gravity are potential fields of both energy and mass. Therefore if gravity, magnetism, energy and mass are manifestations of matter, than indeed there is a unified field theory as envisioned by Albert Einstein. Let us summarize what we have just proposed one more time. Mass of our atom is created at the boundary of negative and positive electric charges when they come into contact with each other, which means that mass is created when opposite electric charges combine. Therefore mass and energy are simply two different phases of the same physical property. If gravity is the manifestation of mass, it is a property of mass therefore it is also a property of energy. Similarly, if magnetism is manifestation of micro gravity therefore of mass, it too is a property of energy. In order to simplify our case, from now on, we will also call electric charges as energy. Based on this assertion, energy and mass are

28

two different phases of matter where magnetism and gravity are manifestations of mass which is a phase of energy. In the end everything is a manifestation of energy. This is the Unified Field Theory in a nut shell.

1.10 Radioactivity

Up until now in this book, we have tried to demonstrate how our atomic model easily explains physical observations such as Brownian Movement, heat transfer and three states of matter, namely solid, liquid and gas states as well as others including particle and wave nature of light and magnetism . However, for our atomic model to be valid, it should also be able to explain radioactivity, radioactive decay, transmutation and chain reactions.

Radioactivity was first observed in 1896 by Antoine Henri Becquerel (1852-1908) while experimenting with naturally fluorescent minerals and photographic plates. During one particular experiment, Antoine Henri Becquerel placed samples of Potassium Uranyl Sulfate, a sulfate of Uranium in mineral form, on photographic plates and wrapped them in black paper to see if the photographic plates were somehow affected by the samples of Uranyl Sulfate. Experiment clearly showed that photographic plates were exposed to a new kind of light which was different from the X-Rays discovered in 1985 by Wilhelm Conrad Röntgen (1845-1923). Becquerel asserted that this new light was in fact was a form of radiation, a new form of energy emitted by the Potassium Uranyl Sulfate. This new form of radiation was later named "Radioactivity" by Marie Currie (1867-1934) while working with her husband Pierre Currie (1859-1906) on the subject.

We are not going into details of radioactivity here since it has been discussed in great detail by many scientists and in academic papers and publications. Our purpose here is to find out if our atom can explain radioactivity at the atomic level for all practical purposes presented in this book.

It is clear from this line of reasoning that transition from an atom of one element to an atom of another element involves a change in its E/M ratio. Consequently our atom can have infinite numbers of isotopes when its mass and energy ratio is somewhat altered. Hence the difference between Uranium$_{238}$ and Uranium$_{234}$ is in their atomic mass as well as in their energy levels.

If our atom has an inherent imbalance between its negative and positive electric charges, it becomes unstable and therefore radioactive. When radioactive, our atom tries to regain its atomic balance by converting some of its mass into its constituents, namely positive and negative electric charges and releasing them to reach a new equilibrium between its electric charges and mass. This process corresponds to transmutation. The release of energy can be in form of energy or mass until a desired amount of energy or mass is released. As a result, our uranium atom can go through a number of stages to become an atom of iron. When our atom reaches a higher level of imbalance between its positive and negative charges and its mass, it can split to form new elements to regain its balance. It is therefore feasible to assert that an element with a higher atomic mass has higher energy levels because when its original equilibrium between its mass and energy is lost, it has more mass to convert into energy.

When its mass dissolves into its components of positive and negative charges, our atom releases excess positive and negative electric charges in order to regain its energy and mass balance. These released excess energy is then in a position to affect the energy levels of other near-by atoms. This process corresponds to nuclear chain reactions. This reasoning leads to a significant conclusion. Nuclear reactions can happen regardless of the critical mass of fissile material involved. What is required is the temperature levels high enough to allow mass of the atomic shell to dissolve into its constituents of electric charges either by pressing them together (fusion) or elevating their temperature to extremely high levels (fission) by external means.

In nature, nuclear reactions happen all the time. Pitchblende, a uranium mineral, releases energy continuously albeit in a much slower pace because of the presence of other non-fissile materials in the mineral. We can observe this release of energy since pitchblende glows under ultraviolet light in a dark room at room temperature. However, non-fissile materials in uranium minerals absorbs some of the energy released therefore preventing a chain reactions to happen. Consequently, in order to release energy at much higher rates by means of chain reactions, uranium or any other fissile material must be refined to higher levels of concentration.

1.11 Atoms at the Speed of Light

So far, in all of our deliberations, we have focused on atoms that are stationary and are near the surface of the Earth. Now, we will take our considerations one step further by asking what happens to our atom if it is accelerated to extremely high speeds, speeds that are in the realm of speed of light. This is extremely important because it explains what happens in a Particle Accelerator or a Hadron Collider. As our atom is accelerated by means of an alternating magnetic force field, it gains speed and therefore energy. We have already established that as its energy increases by which we mean it frequency and amplitude, our atom has less mass hence its E/M ratio has been altered. At a final terminal velocity, our atom becomes all energy and no mass. This means that only energy can travel at extremely high speeds and it is not possible particles with mass to travel at such speeds. This is in total contradiction with our experiments with Particle Accelerators. Is it possible that what we capture and observe in Particle Accelerators are atoms with extremely high E/M ratios? If accelerating atoms to higher speeds increase their energy levels, can we precisely calculate their mass using c^2 as a conversion constant?

Is it possible for two atoms, one stationary and the other travelling at extremely high speeds, to collide? We know that two atoms with their negative charges outside of their shell will have a repulsive force between them when they are brought to close vicinity of each other. We also know, this repulsive force becomes greater as the distance between them becomes shorter. Since this repulsive force is extremely strong in atomic scale, it prevents these two atoms to collide at ordinary speeds. At extremely high speeds, however, the results can be different since atoms at much higher speeds have more energy and less mass. Just like a glass bottle hitting a wall, it is possible to shutter the accelerating atom when it collides with a stationary atom at extremely high speeds. When that happens, shell of the accelerating atom disintegrates, scattering its mass and electric charges. Since this process is completely random in its nature, we can observe infinite numbers of various energy levels in a Hadron Collider. No wonder, we have been discovering large numbers of subatomic particles every day.

Finally, is there an experiment that can possibly support validity of our atomic model? I believe there is one. The Oil Drop Experiment by Robert Andrews Millikan (1868-1953) can be easily explained by our atomic model. An oil drop made of molecules of atoms with their negatives charges outside of their shells can actually float in a strong electromagnetic field if that field has its negative polarity at the bottom and positive polarity at the top, counter balancing the effect of force of gravity oil drop is subjected to.

1.12 Dark Room Experiment

A photon is said to be the elementary particle of light according to the Standard Model. As such, it is said to be the carrier of electromagnetic force and the source of electromagnetic radiation. Furthermore, photons are said to exhibit wave/particle duality meaning they possess properties of both waves and particles. Our Sun is a source of photons just like any other source of light, including a light bulb when subjected to electric current and a candle when lit up. When we walk into a completely dark room, we can't see anything since there is no light therefore there are no photons to observe. Now, if we lit up a candle in this dark room, suddenly there is light produced by the burning of wax or tallow. We know that wax and tallow are made of molecules which in turn are made up of atoms. It seems these atoms are producing photons but how? If we try to touch the flickering flame of a candle, we find out that it is hot just like a light bulb that has been on for a while. So, clearly heat is involved in the process of producing photons. After all, our Sun has a lot of heat so far as we can tell.

Clearly, atoms in the wax or tallow of a candle and the hydrogen atoms of our Sun as well as atoms in the filament of a light bulb are at the source of the processes that generates photons when subjected to heat. Source of heat can be a chemical reaction in the case of a candle or a nuclear reaction in the case of our Sun or an electric current in the case of a light bulb. Somehow, these atoms are emitting photons when heated. Surprisingly, this phenomenon fits very well into our earlier presentation of our atomic model which changes its frequency and amplitude spectrum when heated. In other words, a heated atom has more energy and less mass therefore it is now repelled, released and emitted while glowing with radiant energy. Therefore photons are nothing other than atoms that are emitted after reaching certain heat level so that their frequencies increase to the levels of frequencies of the visible light. Since visible light is a part of overall electromagnetic spectrum, we can assert that our atoms become both photons and sources of radiation when sufficiently heated.

But what about Cathode Rays? How are they generated in a Crookes Tube? A Crookes Tube, invented by scientists including English Physicist Sir William Crookes (1832 – 1919) between 1869 – 1875, is a glass container with two metal electrodes, a negative electrode called Cathode and a positive electrode called Anode, placed in its opposite ends. Air inside the tube is mostly evacuated in order to lower the atmospheric pressure within. When a high voltage electric current is applied to metal electrodes, a glowing light beam appears between the cathode and the anode of the tube. This light beam was later called a Cathode Ray and the Crookes Tube became known as a Cathode Ray

Tube. But how does our atomic model fits into the creation of Cathode Rays?

When a high voltage electric current applied to the cathode and anode of a Cathode Ray Tube, heat is generated. Heat elevates the frequency and amplitude of the atoms and molecules that make up the Cathode, alters their E/M ratios and with that atoms end up having more energy and lass mass. As a result, they are repelled, freed and emitted from the Cathode to fly toward the Anode since Anode is positively charged. Because heat has also elevated the frequency and amplitude spectra of the atoms emitted from Cathode into the spectrum of the visible light, they glow while flying towards Anode. If Cathode Rays are subjected to a magnetic field they drift towards the positively charged plate since the atoms that glow have their negative charges outside. Clearly, our atoms have been misidentified as electrons, starting a scientific journey in the wrong direction leading to the Standard Model.

If this is the case how do we explain the results of Rutherford's Gold Foil Experiments? Rutherford measured something he thought would best fit to the atomic model of his times, atomic nucleus. If the prevailing atomic model of his times were different, I wonder how his interpretation of his measurements would have been different than the ones he proposed at the time. If Rutherford could have known about the atomic model proposed in this book, would he have proposed that he actually measured the size of an atom? We will never know...

But we know this. At the end of our reasoning in this book so far, we have reached an inescapable conclusion that subatomic particles are in fact variations of a single atomic model and that the single atom might in fact be the smallest indivisible particle of all matter. After nearly 2500 years, we are back to the original idea first proposed by Democritus. We have reached a point in our reasoning that we can no longer support existence of subatomic particles as proposed by the Standard Model. The atomic model we have proposed in this book satisfies all tests we have thrown at it so far. Our assertions are emboldened by the fact that at the time of this writing, it was established beyond any reasonable doubt that the Higgs Boson doesn't have enough mass to validate the Standard Model and this changes everything.

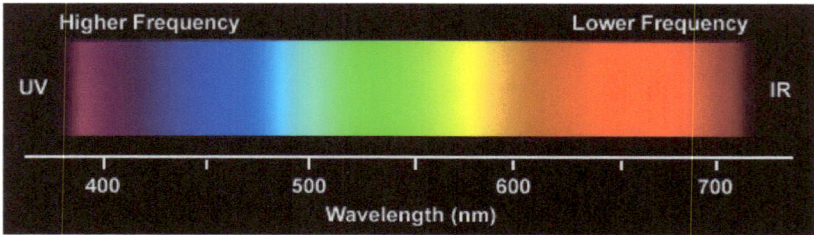

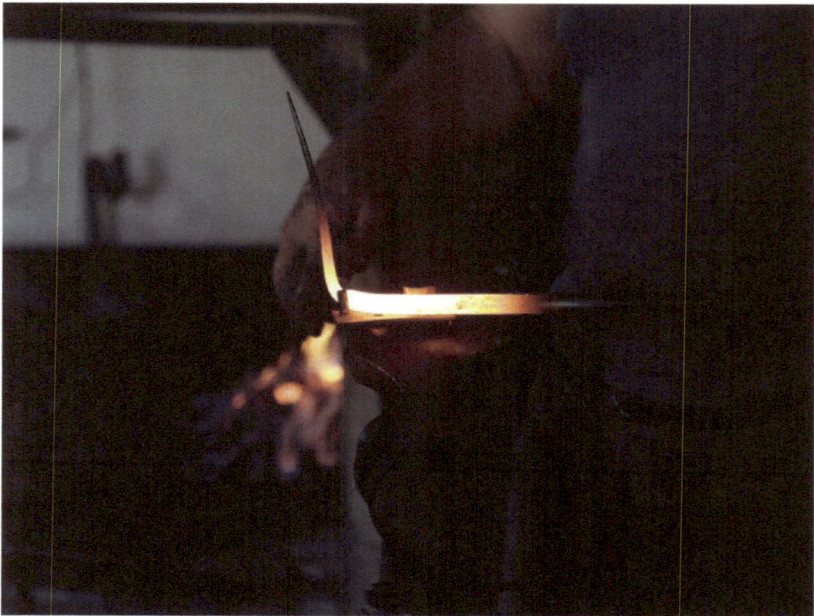

Fig: 1-1a. Black Smith at Work.
Copyright; Flagstaffotos. Reprinted with permission from Henry Firus at
contact@flagstaffotos@com.au

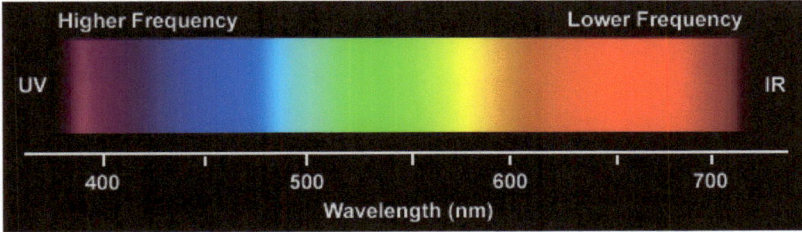

Fig: 1-1b. Another example of blacksmith at work. The characteristics of radiant energy are same with the previous example. It proves that phenomenon of radiant energy does not change from one experiment to another. Reprinted with permission from Henry Firus at contact@flagstaffotos@com.au

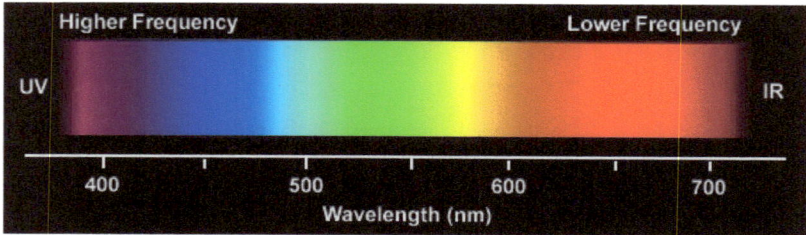

Fig: 1-1c (a). Spectrum of visible light. Source; Wikimedia Commons. Copied under Free Content and Public Domain License.

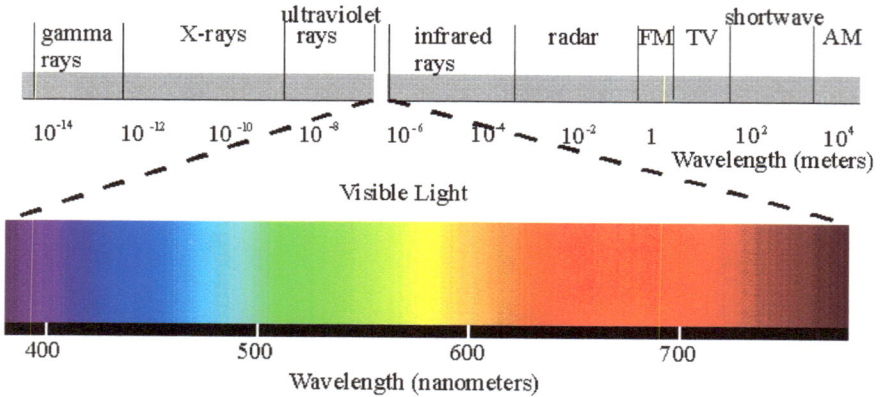

Fig: 1-1c (b). Spectrum of Visible Light as part of Electromagnetic Radiation. Source; Wikimedia Commons. Copied under Free Content and Public Domain License..

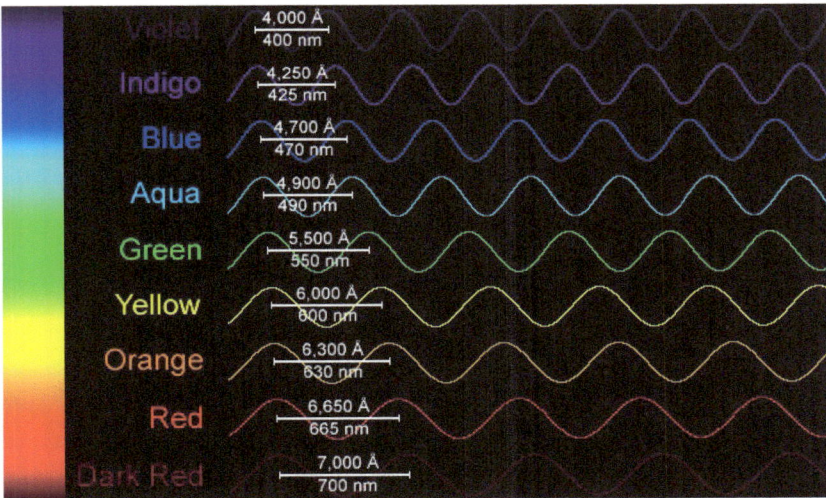

Fig: 1-1c (c). Relation between color and frequency. Source; Wikimedia Commons. Copied under Free Content and Public Domain License.

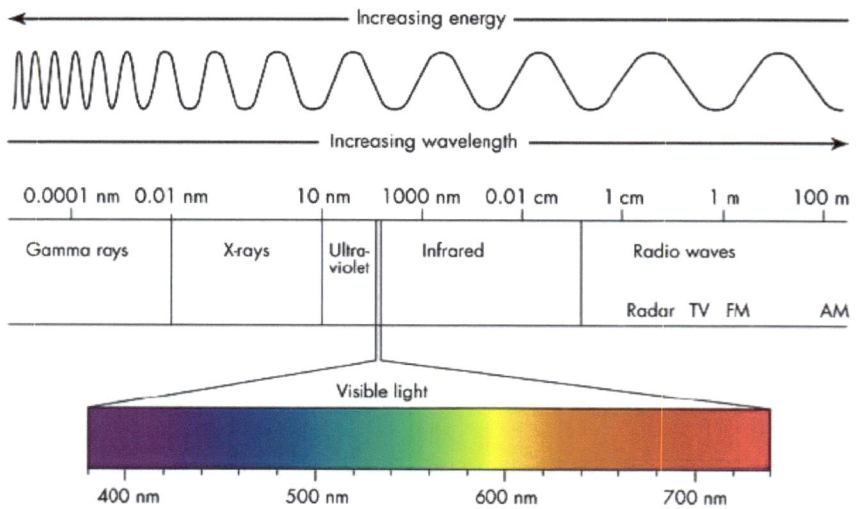

Fig: 1-1c (d). Relation between color, frequency and energy. Source; Wikimedia Commons. Copied under Free Content and Public Domain License.

Fig: 1-1d. Aurora Borealis, also known as Northern Lights, happen when solar winds, which are high energy electromagnetic radiation from the Sun, excite atoms at the upper atmosphere pushing their atomic spectra into the range of visible light in the same manner a microwave oven heats up your food. Photo by United States Air Force employee Senior Airman Joshua Strang. Copied under Free Content and Public Domain License of Wikimedia Commons.

Fig: 1-1e. Another example of Aurora Borealis, Norther lights. Variations in colors are indication of the variations in intensity of the solar winds, a.k.a. solar radiation bursts. Photo copyright; Bjørn Jørgensen. Copied with permission".

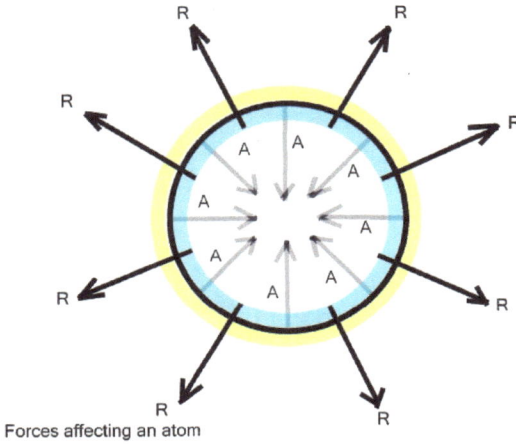
Forces affecting an atom

1-2. Attractive forces, donated by "A", are the forces of microgravity between each and every point of this atom's mass, shown in black and they are trying to collapse the atom to its center while repulsive forces of the positive electric charge, shown in blue, within the atom, donated by "R", are trying to expand the shell. These forces make our atom expand and contract, very much like a pendulum except in 3D. Here negative electric charge is shown in yellow. Please remember that this is a 2D presentation of a 3D object. Illustration by Stephanie C. Houston.

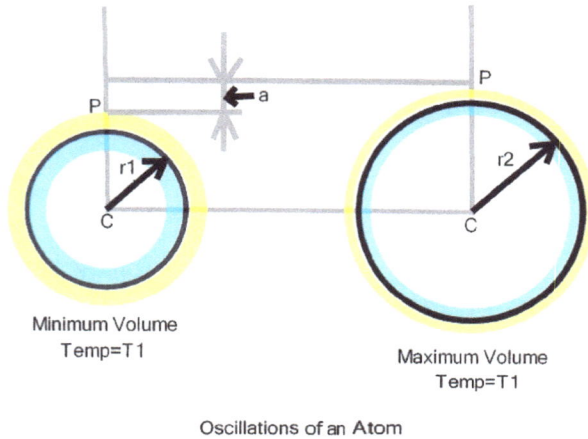

Minimum Volume
Temp=T1

Maximum Volume
Temp=T1

Oscillations of an Atom

Fig: 1-3a. Amplitude "a" is the distance travelled by point "P" on the surface of our atom directly perpendicular to its surface. If we increase the temperature of our atom, "a" increases in all directions. If we plot "a" as a function of time, it is a simple cosine function. Illustration by Stephanie C. Houston.

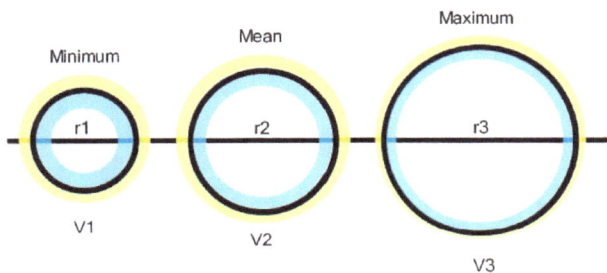

Mean Volumetric Value

Fig: 1-3b. In this illustration, our atom is shown in its minimum, mean and maximum volumes where;

Vmean = (Vmax + Vmin) / 2

and $V_{min} = (4\pi r_1{}^3) / 3$ $V_{mean} = (4\pi r_2{}^3) / 3$ $V_{max} = (4\pi r_3{}^3) / 3$

Illustration by Stephanie C. Houston.

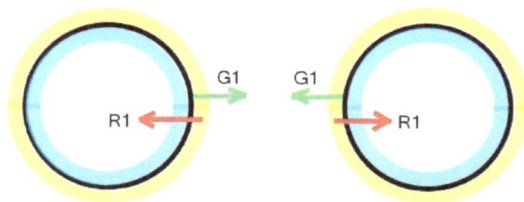

Forces interacting between 2 identical atoms

Fig: 1-4. There are two forces interacting between two identical atoms in close vicinity of each other. Force "R" is the repulsive force between negative electric charges outside of the shell of each atom. This force is trying to increase the distance between these two atoms. On the other hand, force "G" is the attraction caused by the force of micro gravity between masses of their shells. This force is trying to pull these two atoms closer. If the "G" force is greater than the "R" force, these two atoms belong to a solid element. If the "R" force is greater than the "G" force, these two atoms belong to a gaseous element. A near balance between these two forces results in a liquid element. This explains why Mercury is the only liquid element. Illustration by Stephanie C. Houston.

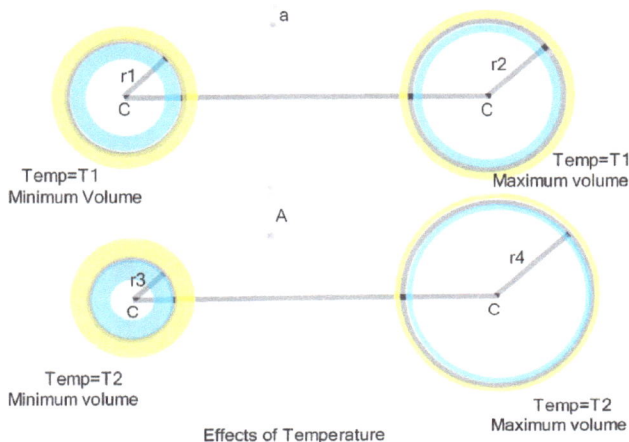

Fig:
1-5 shows the effects of temperature variation on our atomic model. Here, $T_1+\Delta T=T_2$ Increasing temperature from T_1 to T_2 increases the amplitude from a to A. It also increases the frequency from f_1 to f_2 where f_2 is higher than f_1. If we plot frequency and amplitude of the oscillations of our atom, we produce a plot of its atomic spectra. Oscillations of our atomic model are a function of temperature and are probably within Megahertz range at room temperatures. Illustration by Stephanie C. Houston.

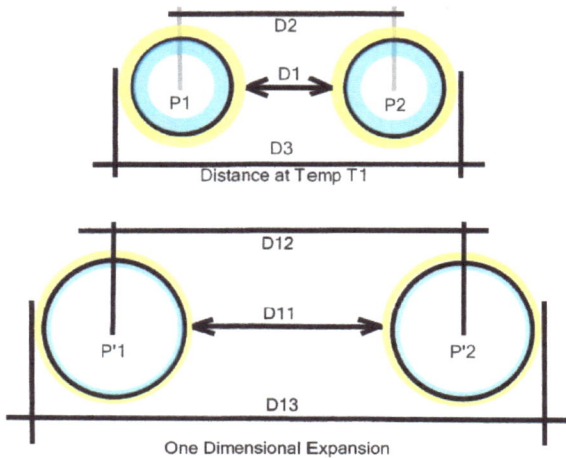

Fig: 1-6. When temperature increases, the distance between the centers of two adjacent atoms also increases in order to allow necessary space between atoms to accommodate their increased amplitude. Here, the distance between P'_1 and P'_2 is greater than the distance between P_1 and P_2. This is called one dimensional expansion. If there is a cluster of atoms involved, this will be a 3 dimensional expansion or volumetric increase.
Illustration by Stephanie C. Houston.

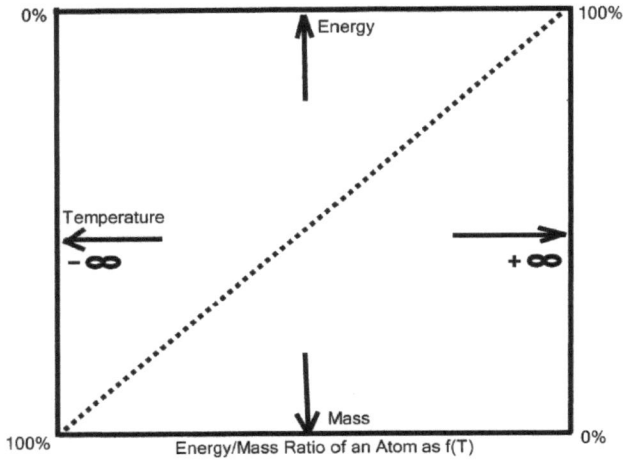

Fig: 1-7. Energy Mass ratio of an atom (E/M) changes with temperature. At extremely high temperature of billions of degrees, our atom is all energy and no mass. At extremely low temperatures, billions degrees below zero, out atom is all mass and no energy. The E/M ratio in this graph is shown with a dotted linear line since it is assumed to be linear. Illustration by Stephanie C. Houston.

Fig: 1-8. Amethytist crystals. Amethytist is a clear purple or violet colored variety of Quartz. Variations in color are due to slight variations in energy levels of the atoms that make up the mineral content. Source; Wikimedia Commons. Copied under Free Content and Public Domain License.

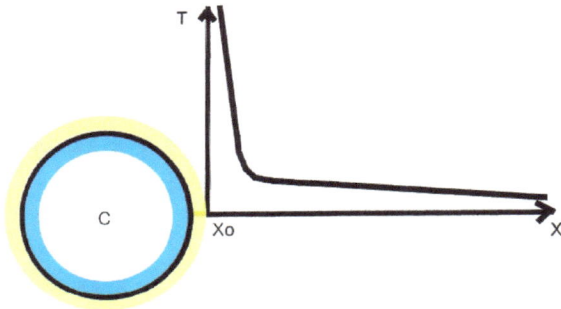

Temperature Gradient of an atom in the direction of x

Fig: 1-9. Temperature gradient of an atom is a simple $1/x^2$ plot, shown in the graph as an approximation, where x is the distance. This gradient is called spherical divergence in a 3 dimensional presentation. Illustration by Stephanie C. Houston.

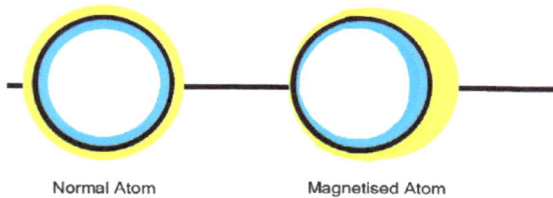

Normal Atom   Magnetised Atom

States of Atom

Fig: 1-10. Comparison of two atoms; the one on the left is not magnetized. The one on the right is. Due to its uneven distribution of electric charges, magnetized atom has an increased attractive force on its left side while it has an increased repulsive force on its right side. When subjected to a magnetic field, a magnetized atom spins to align itself with the magnetic field it is subjected to due to this irregular distribution of its electric charges. Without this property of our atom, computers could not have been possible.
Illustration by Stephanie C. Houston.

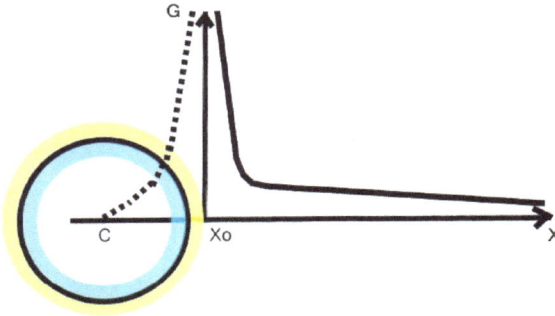

Gravity Gradient of an atom in the direction of x

Fig: 1-11. Gravity gradient of our atomic model in this graph is a simple $1/x^2$ function, where x is the distance. Please note that this gradient is same in all directions creating a 3D volume. If we move our atom back and forth linearly in any direction, it will create gravity waves at the atomic scale in the direction of movement.
Illustration by Stephanie C. Houston.

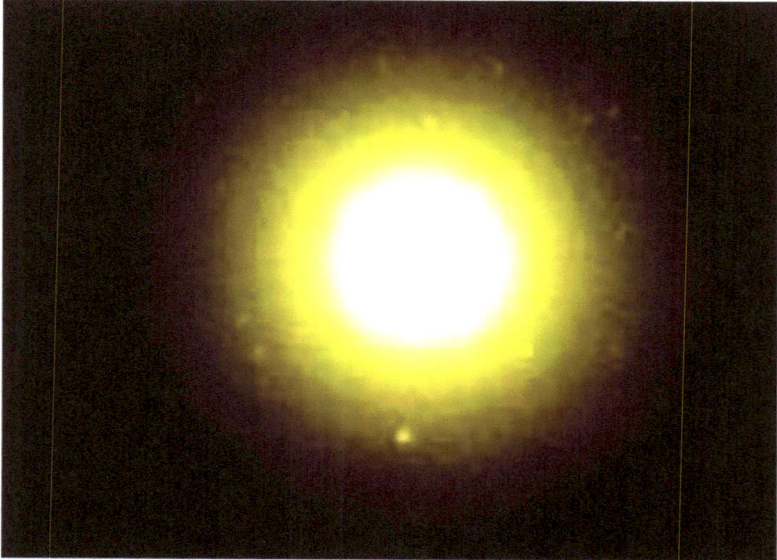

Fig: 1-12. If we could see our atom with our naked eyes, it probably would look like the object in the photo above. We probably couldn't see its volumetric oscillations since they would probably be within the megahertz range. This illustration presents its negative electric charge fairly accurately since its negative charge is stronger near the shell and becomes weaker and eventually disappears with distance at the atomic scale.
Photo montage by Stephanie C. Houston.

CHAPTER 2

SPECTRA OF LIFE

2.1 Definitions

In the previous chapter, we discussed how an atom has its unique frequency and amplitude spectrum. In this chapter, we will look into how that individual spectrum is modified when combined with the spectrum of other atoms within a compound or in a living cell. We know that atoms can combine to create molecules and living organisms. The basic unit of a living organism is a cell and it is made up of atoms of various types or kinds. Consequently, each living cell has its own complex spectrum, which is the summation of the spectra of the atoms that make it up. We will call this Bio-Spectrum when it refers to a single living cell. This is important because it foretells if we are going to get sick when exposed to viruses or bacteria. When a virus or a bacterium enters a human body, its presence is noticed when the bio-spectra of the virus or the bacterium is detected by the immune system cells. This happens because bio-spectra of the viruses and bacteria have an electromagnetic radiation and if any immune system cell is close by, it feels the presence of this electromagnetic radiation as heat. If the virus or the bacterium is dormant, meaning that their electromagnetic radiation or bio-spectra is weak, it may go unnoticed until they becomes active and their electromagnetic radiation or bio-spectra gains strength. When the presence of a virus or a bacterium is detected, immune system cells move in to challenge the virus. Just like the viruses and bacteria, immune system cells have their own electromagnetic radiation or bio-spectra to counter the electromagnetic radiation or bio-spectra of the viruses and bacteria. The outcome of this contest depends on who has the stronger electromagnetic radiation or bio-spectra, in other words, more powerful spectra wins the contest since the stronger electromagnetic radiation disturbs the weak electromagnetic radiation, resulting in the demise of the organism with the weak bio-spectra. This phenomenon has one unavoidable conclusion. A living cell dies when its bio-spectra is altered to some degree. If this sounds like star wars at the nano scale, it is.

Now, let us suppose someone is infected with viruses or bacteria. A medical doctor prescribes a medication. This medication is a chemical compound which in turn is made of various atoms combined in one form or another and as a result, it has a complex spectra. When taken orally or intravenously, it enters the blood stream of the patient and eventually meets the viruses or bacteria. If the medication has more powerful spectra than the bio-spectra of the viruses or bacteria, it interferes with their bio-spectra, which results in their death. If not, it is said that the virus has developed immunity to the medication. This implies that only way to treat a viral or bacterial infection is to find a chemical compound capable of altering the bio-spectra of the invading viruses or bacteria.

Another way to alter the bio-spectra of a virus or a bacterium is to subject them to an electromagnetic radiation, i.e. radiation therapy, where the radiation, which is another way of saying electromagnetic radiation, alters the bio-spectra of viruses and bacteria. It is important to remember that each healthy cell has its own unique bio-spectra. If this bio-spectra of the healthy cell is somehow altered, for example by presence of toxins or carcinogens, the new altered bio-spectra results in a pathological cell. This is how a healthy cell becomes a cancerous cell.

Only way to treat a cancerous cell then is to define the bio-spectra of a similar healthy cell, than calculate an operator, i.e. a radiation therapy, to restore it to its original healthy bio-spectra. Otherwise, the pathological cell dies and so does the patient. One quick note here; based on the considerations we have presented so far, it is safe to assume that any device that emits radiation will affect its surroundings. So if you have a cell phone, I expect that the phone will have an effect on your brain and that effect can be harmful if the radiation of the cell phone is significantly different and stronger than the bio-spectra of your brain. I also suspect that living for long periods of time in close proximity of large arrays of transformers may have a negative effect depending on the distance.

2.2   Fire Flies

During a warm August night at some latitudes, one can observe tiny flying insects in an open field, suddenly emitting a visible red light for a second or two. These are the lights of little beetles, known as fire flies. Fig: 2-1. These insects have the unique ability to alter their own bio-spectra periodically. When they increase the strength of their bio-spectra, the electromagnetic field created by their own cells, becomes strong enough to move into the range of visible light at the red end of the light spectrum. After a second or two, their bio-spectra lose its strength and fall below the spectrum of visible light and into the infrared part of the electromagnetic spectrum. We, humans, have a similar capacity, although it is not within the realm of human control. It is possible that, human immune system cells have the capacity of increasing their bio-spectra, therefore the intensity of its electromagnetic radiation, when confronted by the bio-spectra of a virus or a bacterium. Only possible indication of such an event in human body is that we sometimes feel feverish during an illness, such as a cold or flu. Most of the time after a good sweat, we feel better simply because immune system cells probably overpowered and killed many of the viruses or bacteria in our bodies during a fever.

2.3 Fall Colors

Could all these things we talked about Bio-Spectra be just wishful thinking? Is there a proof in the nature about the existence of bio-spectra? I believe there is and here it is. Fall colors. There is a secret in fall colors that we have for so long failed to understand. When the fall comes, green leaves of trees first turn yellow and then red following the frequency sequence of the visible light spectrum from strong to weak. In other words, the green color of a healthy leaf has stronger frequency and amplitude spectrum than the frequency and amplitude spectrum of the yellow color of a dying leaf. Similarly, the frequency and amplitude spectrum of a yellow leaf is higher than the frequency and amplitude spectrum of a red leaf. The change is always from the green to yellow to red, which is from the higher frequency and amplitude to lower frequency and amplitude in the visible light spectrum. This is because the leaves have bio-spectra and death comes when their bio-spectra falls below the required level of energy. Fig: 2-2. Same phenomenon is behind green tomatoes turning red as they ripen, green bananas turning yellow over days, poinsettia leaves turning red from green when they are kept in the dark for few days as well as aging, poor health and weak metabolisms. Figs: 2-3a & 2-3b. It is feasible to assume that we humans have our unique individual bio-spectra or energy level which would be different for each individual. This individual bio-spectra or energy level is probably strong for some and not so strong for others. It is probably no wonder that some people with weak bio-spectra or low energy levels suffer poor health for all their lives while others with strong bio-spectra or energy levels enjoy a healthy life as long as they live. The fact that we all have different energy levels or bio-spectra is significant. Is it possible then different parts of human body have different energy levels? I suspect the brain and the nervous system have the highest energy levels in human body followed by muscles and internal organs while skeletal frame has probably the lowest energy level.

2.4 DNA

By now, you probably have figured out where we are going with this reasoning. The question, then, becomes how can a single cell at inception becomes a complex organism such as a human body. We all know that our DNA contains genes which have the instructions about how to build a human body from a single cell. In each gene, there are five genetic codes that tell a cell how to modify its frequency and amplitude spectrum in order to morph into a different cell. Let us for a moment assume that, inside each genetic code, there is a cell very much like a light bulb in a dark room photographers use to develop film. Let us also assume that inside of this room is covered with sensitive undeveloped photographic paper. When exposed to light, this photographic paper changes its color to the color of the light it is exposed to. Now, suppose there are five lamps attached to the ceiling of this room, each with its own control switch and each with a different color coating. Let us say that these colors are red, blue, yellow, green and orange. Hypothetically, if we turn the green light on, the photographic paper covering the room will turn green and when developed, it will stay green regardless of what color it is exposed to afterwards. Remember you can expose the film in your camera only once. If we turn the red light on instead, the photographic paper in the room will turn red and will stay red after it is developed. In reality, in human body, instead of switching the light bulb with a specific color on, each gene emits a frequency with a given amplitude within the cell to alter the cell's bio-spectra, sometimes individually, sometimes in a combination of two or more frequencies, modifying the bio-spectra of the cell, therefore changing its chemical and physical properties. When this process is applied before or after a specified number of cell divisions in coordination with the instructions encoded within your DNA, a new type of cell is created each time, in the process building a human body from an initial single cell. These genes that emit a single or a combined frequency for a specific period of time to alter the properties of a living cell without destroying its functionality encourages us to believe that bio-spectra of a cell can be altered without being destroyed. On the other hand, when we cook an egg in a frying pan, we are applying heat, i.e., electromagnetic radiation and the cooked egg now has a different spectra. Although the cells in the cooked egg are no longer alive, the cooked egg still has spectra, although different from the bio-spectra of the egg we originally had. This is important because when we digest food, spectra of food are what we receive in order to keep our bodies nourished. In other words, spectra of food alter and elevate the spectra in our digestive system That brings us to a very interesting conclusion. Some foods have more powerful spectra such as greens than red meat which occupies the lower end of the visible light spectrum. Your grandma was probably right when she suggested that you should eat your greens. Finally, we now know how some substances can cause birth defects when they are present in human body during pregnancy.

While we are still at this subject, let us consider a seed planted in a soil. How does the seed know it is now planted? Only possible explanation I have is that, when we plant a seed, it senses the frequency and amplitude spectrum of the soil in which it is planted, which, in turn initiates the process of cell division and growth. This is why some plants will grow in certain types of soil. If the spectra of the soil do not correlate with the signal the seed is programmed to expect, plant will not grow. One clarification, though, the temperature of the soil the seed is planted into is part of the soil spectra. Some soils are more fertile than others because their spectra correlate better with the spectra of some plants. For example, dark soil, which results from decomposition and disintegration of ancient lava, makes a very fertile soil because it is rich in minerals. Fig: 2-4.

2.5 Brain Waves

Against the background of information we have so far presented, it is safe to assume that our brain has its own bio-spectra which emit its own electromagnetic waves. Medical science calls these waves Brain Waves. Medical science has instruments sensitive enough to measure these electromagnetic brain waves and they are used in diagnosing ailments that affect brain. Occasionally, in a TV broadcast, scans of healthy and not so healthy brains are shown in comparison as diagnostic tools. Color differences in these displays are nothing more than variations of the energy levels of bio-spectra, i.e. electromagnetic radiation of the various parts of an unhealthy and a healthy brain. Brain also operates using electromagnetic signals in order to communicate with the nervous system and the rest of the body. The brain cells that generate these signals are often called neurons and their signals are nothing other than electromagnetic impulses. When you applied a low voltage electric current to a frog leg in the lab to observe its knee jerk reaction, you have actually sent an electromagnetic radiation signal to the nerve cells in the frog leg. This shows that all living organisms are electromagnetic devices and that includes you and me. One interesting foot note here is about telepathy. If brain waves are modified by our thinking, then it is possible to send brain waves as telepathic signals over long distances to communicate with others.

Based on the considerations we have presented so far, we can safely assert that the division between Organic and Inorganic Chemistry is now blurred if not obsolete.

2.6 Colors and Human Eye

We all know that visible light is frequency and amplitude and it is part of the electromagnetic spectrum spanning from Gamma Rays to Radio Waves. When we look at an object, our eyes receive the electromagnetic spectrum reflected from that object as a signal. This electromagnetic signal interacts with the nerve cells in our eye, which in turn is sent out to our brain for processing. The important point here is not really how we see but rather how an object becomes visible to the eye. We all know that in a dark room we can't see because there is no electromagnetic radiation strong enough to interact with the nerve cells in our eye. When the sun rises or when we turn a light switch on in a room, we are generating a three dimensional electromagnetic radiation. In sun's case, its electromagnetic radiation in form of frequency and amplitude reaches the Earth and illuminates everything. To be more specific, when the sun light reflects from a flower, reflected light is altered by the spectra of the flower. If the leaves are green, their bio-spectra alter the reflected light differently than the blooms, which might be red. So the reflected light now carries that information about the leaves and the blossoms to our eyes and to our brain to be seen as green and red. Without this alteration by spectra of living things and substances, our eyes can't separate and identify the colors of the objects we see. Ultraviolet and infrared spectra also have similar affect. For instance, Pitchblende, which is a uranium ore, glows under ultraviolet light whereas under normal day light, no glow is observed because UV light spectrum excites the atoms Pitchblende pushing its spectrum into the range of visible light.

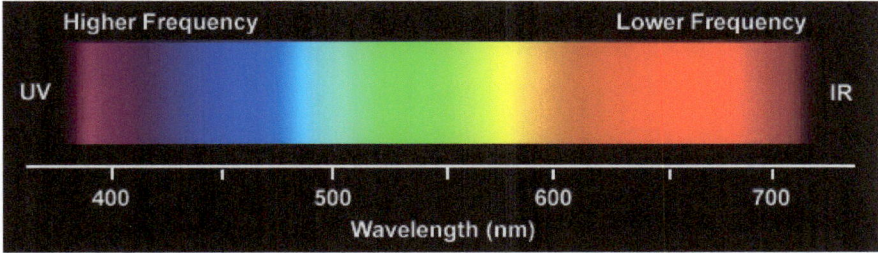

Fig: 2-1. Firefly is a beetle that emits light within the spectrum of visible light that can be seen at night. I suspect, fireflies can perform this feat by flexing their muscles to press their cells closer to increase their bio-spectra into the range of visible light spectrum. Source; Wikimedia Commons. Copied under Free Content and Public Domain License.

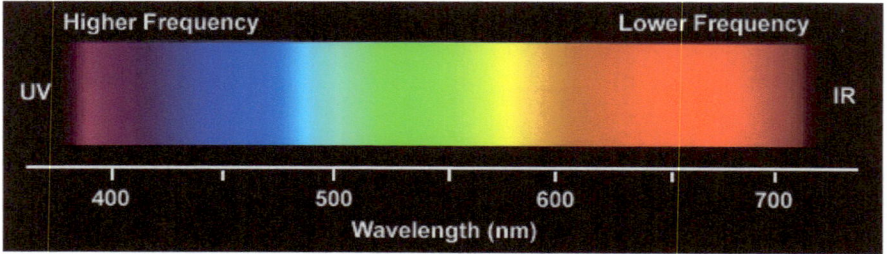

Fig: 2-2. Fall Colors. In the fall, colors of leaves begin to change. Green leaves first turn yellow and then turn red, always in that sequence. This sequence is the exact same sequence of the colors of visible light spectrum, from higher frequency and amplitude of green light to lower frequency and amplitude of red. Photo Courtesy of the author.

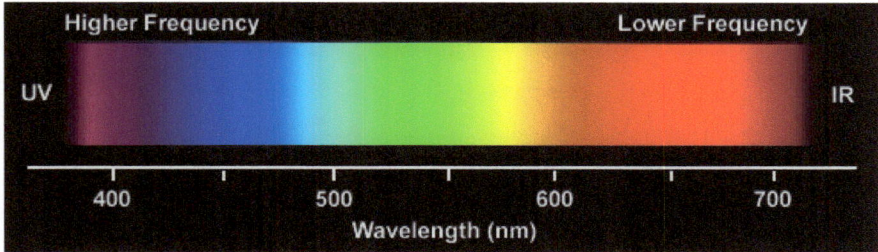

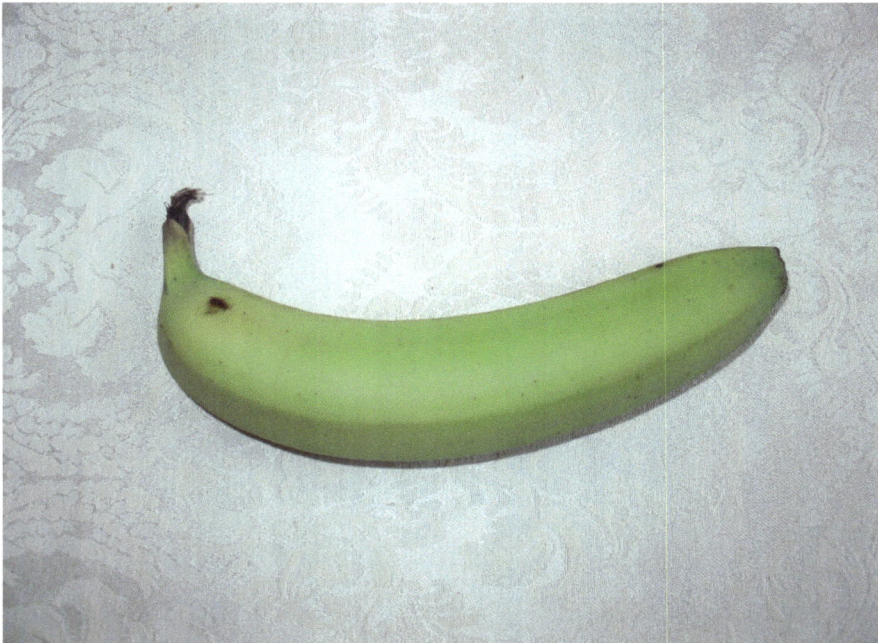

Fig: 2-3a. Green banana. Its green color indicates a higher frequency than the frequency of a ripe yellow banana. Photo courtesy of the author.

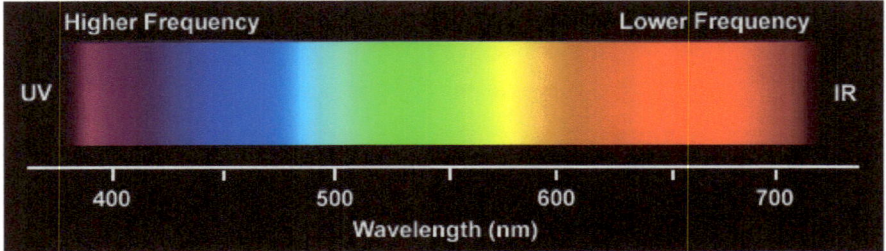

Fig: 2-3b. Yellow banana. This is the same banana in Fig: 2-3a after a few days during which its energy levels have decreased causing its color to change from green to yellow. Photo courtesy of the author.

Fig: 2-4.. Plants thriving on ancient volcanic lava remnant. Mineral rich lava becomes rich dark soil as a result of weathering process. This example is from New Mexico. Photo courtesy of the author

Fig: 2-4a. Spring in Central California. Green grass and shrubs dominate the landscape. Photo courtesy of the author.

Fig: 2-4b. Same landscape in summer. Grass is dead. Yellow color dominates the landscape. Colors have changed from green to yellow, from higher frequency and energy levels to lower frequency and energy levels, once again in the exact sequence of the visible light spectrum. Photo courtesy of the author.

CHAPTER 3

GALAXIES AND TIME

3.1 Introduction

The two fundamental forces of our physical world, namely, the force of gravity and the centrifugal force, seem to be valid within the vast expanses of our observable Universe. The force of gravity is said to be the force of attraction between two masses of any size. The centrifugal force, on the other hand, is defined to be a force that impels a rotating mass away from its rotating axis or center. We are not going into details of these two forces here since they are extensively treated in a number of Physics books of different academic levels.

Based on our current understanding of these forces, we can safely say that in a system of two celestial bodies of significantly different masses, one of which is in orbit around the other, there appears to be a balance between these forces regardless of the distance. However, as the distance between the two increases, mean orbital velocity of the rotating mass decreases and this is well within our expectations considering out present level of knowledge. So if we wish to place two similar satellites into Earth orbits, the one in a lower orbit will have a higher mean orbital velocity than the one in higher orbit.

There are of course no surprises here since planets around the Sun behave exactly in the same manner. While Mercury, the planet nearest to the Sun has a mean orbital velocity of approximately 30 miles per second, Pluto, the 9th planet from the Sun has a mean orbital velocity of about 3 miles per second. On a smaller scale, rings of Saturn act in a similar manner. Inner rings have higher orbital velocities than outer rings. Satellites of Jupiter act in a similar manner if we take into consideration of their relative masses.

On a greater scale, same forces apply to the arms of a Spiral Galaxy. Since these arms are made of masses of significantly different sizes, each and every mass rotating around the central core of galaxy behave in the same manner of a celestial body in orbit around another. The masses in each arm of a spiral galaxy are in orbit around the central mass independently from each other since the gravitational pull of the central mass is the dominating force here although some interaction between near-by masses are expected. The orbital velocity of each mass rotating around the central mass is, in part, a function of its distance from the central mass. Accordingly, masses within the arms rotating near the central mass have higher orbital velocities while the masses rotating at a greater distance have slower orbital velocities.

This implies that the arms of a spiral galaxy are being continuously stretched out as a function of time and distance. This is a far reaching conclusion since it might present clues about the age of a spiral galaxy, therefore affecting our previous perceptions about the age of the Universe.

Let us take this line of reasoning one step further. If the length of arms of a spiral galaxy is a function of its age, then using Galaxy Classification Chart presented by Dr. Timothy Ferris, Fig. 3-1a, we can safely deduce that an Sd galaxy is younger than an Sc galaxy while an Sc is younger than an Sb. Similarly, an Sb is younger than an Sa. We can now say that we have an idea about how a spiral galaxy evolves as a function of time.

We are now ready to consider the form of a spiral galaxy before it becomes an Sd galaxy. In order to do this, however, we must travel back in time. If the arms of a spiral galaxy stretch as a function of time, then, reversing this process, we should reach a form of a spiral galaxy consisting of a central mass with two linear arms pointing straight out and perpendicular to its surface at two opposite points at its equator. This is significant because it implies that spiral arms may have been ejected from the core of the central mass of the spiral galaxy at the very beginning. If this is true, than, a spiral galaxy might have looked like a barred spiral before evolving into an Sd galaxy.

Let us review what we have proposed so far. We have a massive central mass, possibly a massive black hole, whose core is under immense pressure and heat, ejects two columns of overheated plasma (here plasma refers to atoms of 100% energy and no mass) from two opposite points at its equator with the help of the centrifugal force created by its spin around its axis. These two newly ejected plasma columns then slowly begin to evolve to become spiral arms as they begin to cool and convert from plasma to mass as they expand in all directions. At this stage, a spiral galaxy possibly looks like a barred spiral, more specifically like an SBa galaxy. It, then, sequentially evolves into an SBb, SBc and SBd before becoming an Sd galaxy.

But what the future holds for a spiral galaxy? I suspect that as the stretching of spiral arms continue, each arm revolves around the central core multiple times and as a result they begin to fill the once unoccupied space with mass. Consequently, there is now increased gravitational interaction between the adjacent arms. This process begins to break up spiral arms. Cartwheel Galaxy is an example of this stage of breaking up. In the end, some segments of original spiral arms in part form an outer ring while some of the mass fall back at the central core. Once this process is completed, our spiral galaxy is now a ring galaxy. Hoag's Object is an example of a typical Ring Galaxy.

Ring galaxies are the most stable forms of galaxies. However, over billions of years, the ring, converts into 100% mass by further cooling and becomes more susceptible to outside gravitational pulls. This leads to disintegration under the effects of other near-by gravitational forces leaving central core to start all over again to gather new mass as it plunges through the great vastness of space to become another massive black hole in order to start a new spiral galaxy.

3.2 Dark Matter

In previous chapters, we discussed how atoms in extremely cold temperatures become mostly mass and very little energy. When atoms become mostly mass, as in gray matter, or 100% of mass as in dark matter, the gravity fields they create are much stronger allowing their attractive force to extend to reach greater distances in space to pull in more of the freely floating matter in their vicinity allowing the formation of an initial mass. Once formed, this initial mass gets larger and bigger, its density increases several fold, its gravity field becomes stronger and reaches further and the process of gathering dark and gray matter accelerates exponentially. Eventually, this initial mass reaches a point in which it becomes a black hole whose gravity field is so strong that no mass can escape its gravity field. Big black holes are said to have devoured galaxies and star clusters at an alarming rate as they plunge through the Universe.

3.3 Anatomy of a Black Hole

A massive black hole, cold on the outside, can have an extremely hot interior since atoms in its core are pressed against each under the weight of mass above. (Remember the temperature gradient of atoms?) As black hole gets larger, the temperature in its interior eventually reaches millions of degrees. At these very high temperatures, its interior is mostly plasma, another way of saying 100% energy, with pressures reaching millions of pounds per square inch, striking a balance between the gravity force that is trying to collapse the black hole to its center and the pressure created by the electric charges that are trying to expand. At this stage, a black hole has two specific characteristics. First of these, of course, is its immense gravity field. But it also has an immense field of electric charge because of the electric charges trapped inside at extremely high temperatures create a charge field outside. So if the path of a light beam is bent in the near vicinity of a black hole, it is probably because of this charge field instead of its strong gravity presence since we assume that photons have no mass and therefore they are not affected by any gravity field regardless of how immense that gravity field might be. (The other probability is that photons might have an extremely small mass therefore are affected by the immense gravity field of a back hole)

As the black hole grows bigger and bigger, more of it mass transforms into electric charge leaving a relatively thinner shell, proportionally speaking, to contain a steadily increasing pressure inside. At one point after reaching a critical inner pressure level, a black hole either explodes to become a Super Nova, if the black hole has no spin or a Galaxy, if it has spin.

If we could peer into a black hole and see it inner structure, we would observe a central spherical core of plasma of atoms of 100% energy. Immediately next to the core, we will find a transition zone of gray matter in which atoms are differentiated based on their E/M ratios. While those atoms with high E/M ratio are near the core, others with low E/M ratio are at an increasing distance as a function of their temperature.

Finally, dark matter of atoms that are 100% mass envelopes the entire core and sphere of gray matter to complete the structure of a black hole. Because of this reasoning, we must propose that all black holes are spherical in shape and that is in line with other spherical celestial objects we have observed so far.

Now, we must remember that the dark matter that envelopes the entire structure of a black hole has an immense density and this density increases as the black hole gains mass and contracts. Furthermore, as the black hole gains mass and contracts, its spin proportionally increases until it reaches a critical angular speed. At this extremely high angular velocity, the mass at the equator of a black hole begins to feel the effects of this centrifugal force. Accordingly, the gravity force that is pulling back the dark matter of its shell is now weakened by the centrifugal force that is trying to impel the same dark matter. Once the balance between these two forces changes in favor of the centrifugal force, a black hole begins to eject extremely hot gray matter and plasma straight out at its equator.

3.4 Spiral Arms

In the previous paragraph, we said that, for a black hole to become a galaxy, it must have spin. The reason for that is simple. Rotation creates centrifugal force and centrifugal force helps the pressure inside the black hole to find a weak spot to punch a hole at its equator. In other words, once the pressure inside the black hole finds a weak point at the equator of the black hole because of its high angular speed, it pierces through the shell and ejects a column or two or more of hot atoms with no mass straight into the space. The ejection of extremely hot atoms of pure electric charges continues from one or more holes as the black hole continues to collapse on itself, cooling its hot interior in the process while providing additional pressure to help ejection process to continue.

(Think of squeezing a lemon to get the juice out) Once the pressure inside the black hole falls below the pressure levels necessary to eject hot atomic column, ejection stops but the black hole continues to collapse on itself until it can no longer do so and its spin, increasingly faster as its mass moves closer into its center.

Depending upon the initial pressure levels of hot atoms at or near the core of the black hole, amount and the speed of ejected material and the angular speed of the black hole during ejection, dictates the form of spiral arms of the galaxy. This, in a nut shell, is the birth of a spiral galaxy. It is important to remember that, during ejection, hot atoms with no mass, freed from being under immense pressure in the core, begin a process of rapid cooling and acquiring a mass while still travelling outward. However, after a rapid initial cooling and acquiring mass, the cooling process slows down and as a result, galaxy arms are composed of atoms of various rates of mass and energy. We have here clarified once for all how a massive black hole with its immense gravity field allows ejection of spiral arms and formation of spiral arms despite the fact that no light is said to escape its overpowering gravity field.

It is important to remember that after the ejection of spiral arms stop, black hole at the center of galaxy still has a very hot core and therefore has a very strong electric charge field which excites the atoms engulfing the black hole, allowing them to emit radiation. This is why a black hole at the center of a galaxy shines brightly when its radiation levels fall into the visible range of light spectrum at a distance on cosmic scale.

One unique property of spiral arms at this stage is that they are mostly made up of atoms with different E/M ratios. This is critical since as the cooling continues, clusters of mass with different E/M ratios begin to form new gatherings of mass. This is how stars and planets are created when this newly created masses increase in size and develop their own hot interiors. Our Sun is a product of this process. Stars in intergalactic space are examples of the same process. Without existence of the temperature gradient of our atom, galaxies and stars could not have been possible.

One final but very important note about the galaxy arms... In much younger galaxies whose spiral arms are mostly made up with matter of high E/M ratios, galaxy arms are affected by both the gravitational field created by the mass of the black hole but also the electric charge field created by the plasma in the hot core of the black hole. As a spiral galaxy matures however the influence of the electric charge field diminishes because galaxy arms are now mostly matter and therefore contain a lesser amount of matter with high E/M ratio. Similarly, arms closer to the central mass are influenced to a greater degree by the electric field created due to existence of hot plasma in the core of the black hole. This explains why some spirals have bars when they are very young and have massive amounts of matter with high E/M ratio. The central bar in a spiral galaxy however eventually disappears when high E/M ratio of matter near the central mass slowly drops to low E/M ratio therefore decreasing the effects of charge field created by the plasma in the core of black hole. As a result, it is safe to propose that spiral arms of a galaxy will not only obey the laws of gravity, but to some degree to the laws of potential fields of electric charges as a function of the age of a spiral galaxy.

Let us summarize what we have proposed so far. Our spiral galaxy begins life as a massive black hole spinning around its North/South axis with an increasingly greater angular velocity. As a result, the massive black hole in our illustration is more of an ellipsoid than a sphere. (Our Earth is also an ellipsoid, albeit slightly but not perfectly. In the field of Geodesy, its shape is often referred to be a Geoid).

This author suspects that Elliptic and SO galaxies do have massive central masses spinning at extremely high angular velocities which transform their shapes from a sphere to an ellipsoid of various degrees. As a result, ejected hot matter fails to form individual arms but rather engulfs the central core. In other words, in the opinion of this author, elliptic galaxies are failed spirals. In the case of an SBO Galaxy, some of the ejected material forms a Saturn like ring.

Once inside pressure with the help of centrifugal force created by its axial spin helps to eject, in this case, two columns of overheated matter of high energy atoms forming the two nearly spiral arms directly pointing away from the central mass.

As the cooling begins, atoms of the elected mass acquire a higher percentage of mass. As a result the two ejected arms begin to feel the effects of the gravity field of the central mass. Consequently, more developed spiral arms begin to form. A young spiral galaxy has more energy and less mass while a mature spiral galaxy has more mass and less energy.

Our Milky Way Galaxy is in transition from stage g to stage h therefore is a mature spiral galaxy in the very early stages of splitting and breaking up its spiral arms. Hoag's Object on the other hand is in stage k while SN1987A is in stage l. Please remember that in very early stages of galaxy development, the charge field of plasma in the core of central mass has a greater influence in the shape of spiral arms near the core because the E/M ratio of the matter that makes up the arms is relatively high. As the galaxy gets older this E/M ratio becomes relatively lower therefore the force of gravity has a significantly more effect on the shape of spiral arms.

Elliptical Galaxies

E   E4   F5

SO Galaxies

SO   SBO

Spiral Galaxies

Sa   Sb   Sc   Sd

Barred Spiral Galaxies

SBa   SBb   SBc   SBd

Fig 3-1a Galaxy Classification

Illustration from *Galaxies,* by Timothy Ferris

Copyright © 1980 by Timothy Ferris

Reprinted with permission of Sierra Club Books

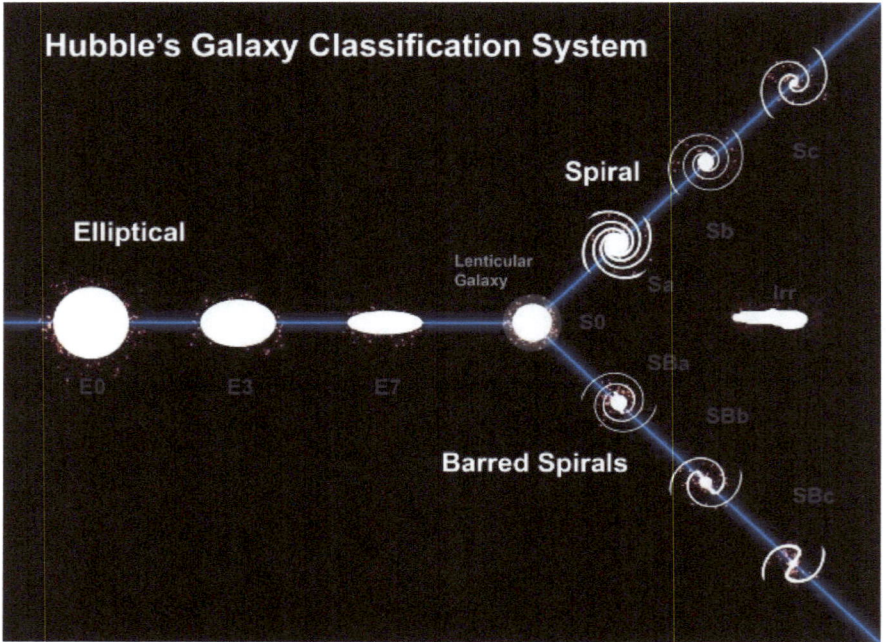

Fig 3-1b, A presentation of Edwin Hubble's Galaxy Classification.

Source; Wikimedia Commons. Copied under Free Content and Public Domain License.

Fig: 3-1c. Proposed birth, evolution and death of spiral galaxies. Courtesy of the author.

Fig; 3-2. NGC 4650. A very young barred spiral galaxy. It has ejected its plasma from the central core but no spiral arms have been fully formed yet. Image courtesy of NASA/STScI.

Fig: 3-3. NGC1300 is classifies as a Barred Spiral. Its bar is the remnant of recently ejected plasma while spiral arms are just beginning to form. This is a very young spiral galaxy. Image courtesy of NASA/STScI.

Fig: 3-4. NGC 1566. A very young spiral galaxy somewhat more mature than NGC1300 since its bar has evolved into spiral arms to a greater degree. Image courtesy of NASA/STScI.

Fig 3-5. NGC 1365 is another very young spiral galaxy. Differences in the shapes of spiral arms are the result in variations in angular velocity of the central core, E/M ratio of ejected atoms and plasma, ejection as well as cooling rates. Image courtesy of NASA/STScI.

Fig: 3-6. Spiral galaxy NGC4639 has a brightly shining core which is a massive black hole engulfed in newly formed atoms with high E/M ratios that create an intense electromagnetic radiation that is visible at great distances. Image courtesy of NASA/STScI.

Fog 3-7. NGC6946 is a young spiral galaxy however it is more mature than NGC1566 and NGC1365. Image courtesy of NASA/SRScI.

Fig 3-8. NGC 3310 is also a very young galaxy with an age comparable to NGC 1566. Image courtesy of NASA/STScI.

Fig: 3-9. Spiral galaxy NGC5457. A young vibrant galaxy estimated to be twice as big as our Milky Way Galaxy. Spiral arms are vibrant with high energy content and are host to a great number of recently formed stars. Image courtesy of NASA/STScI.

Fig: 3-10. NGC6946 is a vibrant young adult spiral galaxy however more mature than NGC 1566. Image courtesy of NASA/STScI.

Fig: 3-11. NGC 5364 is another young adult galaxy whose spiral arms have already encircled the core more than once. Image courtesy of NASA/STScI.

Fig: 3-12. NGC3031. A young adult spiral galaxy, comparable in age to NGC 5364. Image cortesy of NASA/STScI.

Fig: 3-13. NGC628 is a beautiful example of a fully formed young galaxy whose spiral arms are full of newly formed stars. Image courtesy of NASA/STScI.

Fig: 3-14. NGC1232 is a fully developed spiral galaxy with well defined spiral arms brimming with newly formed stars some of which may be comparable to our Sun. NGC 1232 is clearly more developed than NGC628. Image courtesy NASA/STScI.

Fig: 3-15. Spiral Galaxy NGC3370. A vibrant mature galaxy surrounded by energy rich spiral arms, featuring young stars whose nuclear chain reactions are in full swing. Its spiral arms have encircled the central core more than once. Another distant spiral galaxy can be seen edge on at 5 o'clock of the image. Image courtesy of NASA/STScI.

Fig: 3-16. NGC2841 is a fully developed mature spiral galaxy whose arms encircled the central core several times. Please note that as arms continue to encircle central core, they will end up getting closer to each other allowing a greater degree of gravitational interaction between arms which eventually will lead to disintegration of spiral arms. Image courtesy of NASA/STScI.

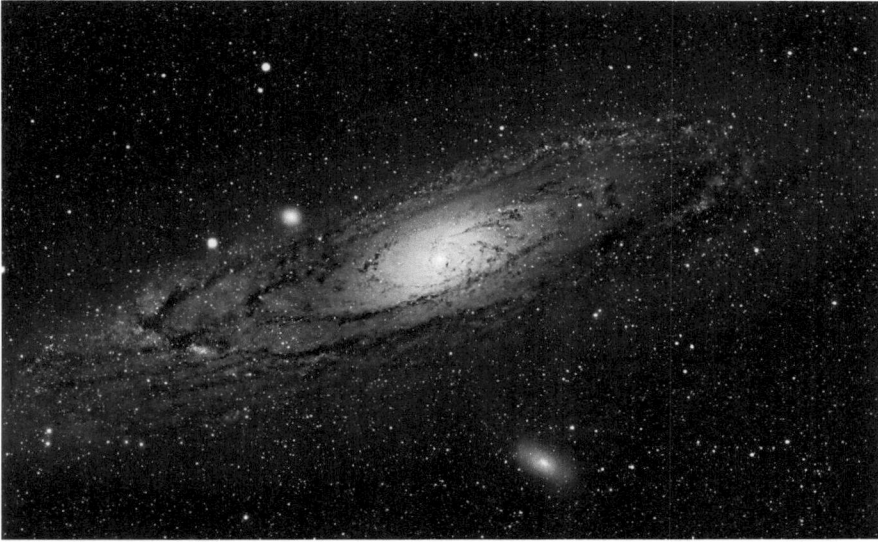

Fig: 3-17. NGC224, Andromeda Galaxy, seen here nearly edge on, is mature spiral galaxy with a binary system of black holes in its center. It is possible that larger of the two black holes reached a critical mass and started ejecting spirals arms while the smaller black hole was still dormant. The fact that these two black holes have not collided despite their immense gravity field may be an indication that black holes have intense energy fields surrounding them which creates a repulsive force that keeps them apart. Andromeda is the nearest spiral galaxy to our own Milky Way. Image courtesy of NASA/STScI.

Fig: 3-18. Central core of a spiral galaxy is seen near edge on. This is a mature galaxy with most of its energy has converted into mass. The black hole in its center glows because it is engulfed by atoms with high E/M ratios which are kept in place by its immense gravity field. It is important to remember that these atoms are constantly excited under the influence of the electric field of the hot atoms in the hot core of the black hole. Image courtesy of NASA/STScI.

Fig: 3-19. Dusty Spiral Galaxy NGC 4414. This galaxy is probably formed by a black hole with a high angular velocity and a slow ejection rate of spiral arms. Image courtesy of NASA/STScI.

Fig: 3-20. Cartwheel Galaxy is an example of an advanced interim stage between a spiral and ring galaxy. As spiral arms of a galaxy breaks up they form an outer ring while some of the mass from spiral arms fall back to central core. A competing theory proposes that Cartwheel Galaxy might have been formed when a familiar structure of a spiral galaxy is disturbed by a near passage of another celestial object, perhaps another galaxy. Image courtesy of NASA/STScI.

Fig: 3-21. Hoag's Object is a fully formed young ring galaxy. Remnants of segments of the original spiral arms are recognizable in the freshly formed ring. Other ring galaxies are also observable at a distance behind the Hoag's Object in this photo. I have observed 4 more ring galaxies. Can you find them? Hint: One of them is at 1 o'clock position just inside its ring. Look for a faint red spot with a ring around it. Image courtesy of NASA/STScI.

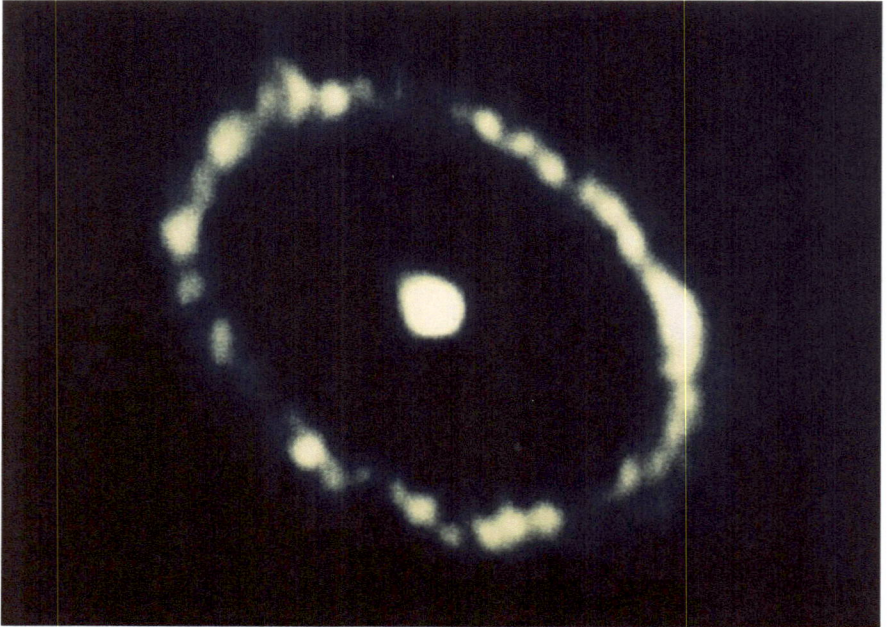

Fig: 3-22. SN 1987A. This image is said to be a supernova remnant. However, the presence of its core suggests that this may indeed be a much older ring galaxy judging from the advanced stage of deterioration of its ring. Furthermore, image of a supernova explosion is three dimensional while this image appears to belong to a two dimensional object. Image courtesy of NASA/STScI.

Fig: 3-23. If we could travel out of our home galaxy to a point directly above its central mass, Milky Way Galaxy, our home galaxy, would probably look like this image here. Our Sun is embedded somewhere in its Orion Arm. Source; Wikimedia Commons. Copied under Free Content and Public Domain License.

Fig: 3-24. A concept presentation by NASA of our Milky Way Galaxy seen near edge on as it could be seen by a Cosmic traveler. Image Courtesy of NASA.

Fig: 3-25. Looking at our Milky Way Galaxy from inside out as we normally do at night in this what appears to be a long exposure photograph. Source; Wikimedia Commons. Copied under Free Content and Public Domain License.

Fig: 3-26. A typical Spiral Galaxy seen here in bird's eye view and edge on. Presence of halo might be an indication of presence of a charge field as excited by the hot core within the central mass. Source; Wikimedia Commons. Copied under Free Content and Public Domain License.

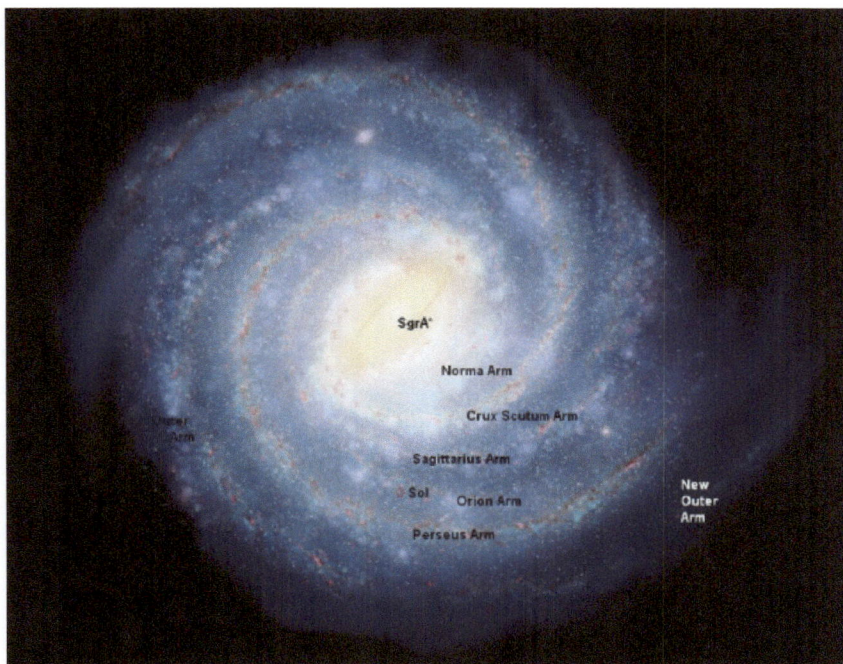

Fig: 3-27. Conceptual presentation of the spiral arms of our Milky Way Galaxy, namely Perseus Arm, Orion Arm, Sagittarius Arm, Cruz Scutum Arm and Norma Arm. A new outer arm is also shown. Word Sol presumably indicates the location of our Sun in the Orion Arm. Source; Wikimedia Commons. Copied under Free Content and Public Domain License.

Fig: 3-28. NGC325-G004 Elliptic Galaxy. Elliptic galaxies lack spiral arms and the plasma surrounding the central mass is highly energized. It is possible that elliptic galaxies are failed spirals.

CHAPTER 4

FOOT PRINTS OF A PLANETARY SYSTEM

4.1 Definitions.

Let us, for a moment, suppose that we have two spherical celestial objects of same mass and they are in an orbit around a third but much larger central mass, following each other fairly closely. As we would expect, in addition to the gravity pull between the central mass and these two objects respectively, there is also a gravity pull between these two objects. If this wasn't the case, these two objects will have the same orbital velocity due to their equal mass and distance from the central mass and therefore will follow each other at a constant distance in their shared orbit. However, in reality, there is indeed a gravity pull between these two and as a result the one in front is being slowed down while the one following is gaining speed. The object that is being slowing down falls into a relatively lower orbit closer the central mass while the object gaining speed climbs up to a relatively higher orbit. Soon the object with slower speed is in an orbit below the object with higher orbital velocity. As the faster object in higher orbit passes over the slower object in lower orbit, the gravity pull of the higher object is now pulling lower object higher. As the passing action is completed, the same gravity pull is now causing slower object to speed up and climb up to a higher orbit while the faster object to slow down and fall into a lower orbit. These interactions continue until both objects, once again, have the same orbital velocity and are in the same orbit again, only to repeat the cycle over and over. Figs: 4-1, 4-2, 4-3, 4-4 and 4-5.

Therefore it is fair to say that these two objects are in an orbit with same mean distance from the central mass and have a phase difference of 180 degrees between their nearest and farthest points from the central mass to form their individual trajectories that are braided together. Since these two objects are equal in mass, the loops of the braid are equal in length and width and display symmetry. As for the orbital velocity of these two objects, we can say that their orbital velocity changes between a minimum and a maximum, again with a phase difference of 180 degrees. Furthermore, we must now speak of a mean orbital velocity, which is same for both.

If we have a situation, in which, one of the orbiting masses is significantly larger than the other, very much like the Sun and the Earth orbiting central mass of our Milky Way Galaxy, then the orbital trajectory of the larger mass will be almost straight for all practical purposes while the orbital trajectory of the smaller mass will form the loops of a braid. These loops will be asymmetrical in shape and their lengths will differ since a short one will be followed by a long one in a pattern repeated for every other loop. In addition, the mean orbital velocity of the smaller object will be equal to the mean orbital velocity of the larger object. Its instantaneous orbital velocity on the other hand will depend on its position with respect to the larger object and will continuously change between a minimum and a maximum. Hence, the Earth's orbital velocity continuously changes as its travels around the central core of our galaxy while it forms braided asymmetric loops with the Sun's orbit. As a result, the Earth's orbit around the Sun is an ellipsoid if plotted with the position of Sun as being stationary. Similarly, all planets in our Solar System have their own braided orbital trajectory with the Sun, only difference being the length of the loops which increase with the distance from the Sun. Figs: 4-6, 4-7 and 4-8.

4.2 A Star Is Born

Based on the information we have so far presented in this book, our Sun probably started out as a very small lump of mass within the inner fringes of the Orion arm of our Milky Way Galaxy during the early stages of the galaxy formation. At this very early stage, it was mostly made up of atoms with different E/M ratios and as a result, it was mostly a mixture of plasma. As its cooling continued and more of its energy transformed into mass, it continued to capture more atoms and particles floating nearby while undergoing a process of differentiation, allowing heavier atoms to form a spinning central core while its gaseous material gathering around this central core to form the Sun's body. This body of gaseous material also began to spin as it began to contract under the influence of developing gravity and radiation volume of the core, very much like the arms of a spiral galaxy, spinning faster near the core and slower at the surface.

As the Sun's central core gathered more mass, it began to exert an ever increasing force of gravity pull on the atoms near its core, pressing them against the surface of the core as well as against each other, allowing their temperature to rise to reach millions of degrees. At these very high temperatures, these atoms are all energy and no mass, therefore moving away from the central core of the Sun leaving a fewer number of atoms within this sphere of critical volume, easing the pressure on the atoms and allowing temperatures to drop considerably. Once the temperatures drop, more atoms are pulled back in the critical volume sphere by the force of gravity, once again increasing pressures and temperatures therefore allowing a new cycle of temperature increase and expansion to begin. It is clear from this line of thinking that our Sun is very much like a self-regulated nuclear reactor, its volume expanding and contracting to radiate energy in discrete amounts. Since nuclear reactions near the core generate extreme amounts of heat, this heat is transferred to the surface by radiation allowing formation of Coronal Loops and Solar Prominences. Coronal Loops are atoms that are all energy at the beginning, therefore rising away from the Sun's surface despite its immense gravity pull, cooling while moving away, gaining mass as its cooling continues and falling back to the surface of the Sun under the influence of its gravity once its cooling allows the formation its atomic mass. Figs: 4-9a, 4-9b and 4-9c.

We can deduce several conclusions from what we have presented about our Sun so far. All stars, small or large, have massive central cores. Without it, stars cannot ignite. By this token, Jupiter is probably is a failed star, unable to create enough pressures and temperatures needed to start its nuclear furnace. Second, the cores of stars as well as their gaseous bodies spin. Third, stars generate massive gravity and radiation fields. Fourth, the temperature at the surface of the stars is not uniform. This is probably because the nuclear reactions near the core are random in nature.

As our Sun continues to shine, its core continues to gain mass, increasing in volume and spinning faster and faster. In contrast, its gaseous body around the central core becomes thinner, oscillating with increased frequency as the time goes by. Near the end of its active life, our Sun becomes a Pulsar, emitting large amounts of energy in short bursts of ever increasing frequency. Once all of its fuel depleted, our Sun becomes a source of radio waves. At the very end, our Sun ends up being Dark matter.

Let us for a moment suppose that, we lift off from the Earth in an imaginary space ship. Once in orbit, high above the Earth, we see Our Sun's light as near white light. If we leave this orbit and travel towards Sun, this light becomes ultraviolet light as a result we can no longer see our Sun. We can detect its presence by measuring its radiation, but we can no longer see it. On the other hand if we travel away from the Sun in this imaginary space ship, the color of our sun will first appear increasingly yellowish, slowly turning orange and eventually appearing red as it becomes smaller and smaller as the distance from the Sun increases. Beyond that we can no longer see our Sun, but we can detect its presence by measuring its infrared, microwave and radio waves depending on our distance from it.

If we are traveling in space in this imaginary space ship and find ourselves approaching a star, we first detect its presence by its radio waves. Closer, these radio waves become microwaves first and infrared waves as we move in closer. Then suddenly we see the star as a source of red light. As we continue to approach, the star becomes yellow in color, and then becomes green, followed by blue and violet. Closer, once again, we can no longer see the star, but we can detect its radiation, first as ultraviolet radiation, then as X-Rays and finally as Gamma Rays. I am guessing that, there might be a planet in the vast expanses of Universe whose inhabitants wake up to the green light of their sun in their morning. Fig: 4-10a, b, c, d and e.

Finally, we asserted that the heat distribution on the surface of our Sun is not uniform. Sun Spots may be areas of electromagnetic radiation above range of visible light consequently appearing as dark spots. Judging by the variations in Sun Spot activity, we can assume that the Sun's radiation output may not be constant over time. Astronomer Carl Sagan (1934-1996) had brought to the attention of scientific community that the disappearance the Sun Spots for a period of 75 years in 1600s coincided with a mini ice age that was observed in Europe during the same period. In fact, variations in Sun's long term radiation activity may have been responsible for the mass extinction of dinosaurs and other forms of early fauna and flora, formation of metamorphic rocks and red beds as well as the sea level changes throughout the geologic ages. Only the Sun is capable of exerting such dramatic changes on planet Earth because of its energy potential and close proximity.

4.3 Planetary Systems

Our planet, like all other planets whose orbital planes lie within the equatorial plane of the Sun, may have been a part of the Sun's composition before it was separated by a catastrophic event. In order to understand this separation, however, we will have to visit the interior of our Sun again. We must remember that the core of our Sun is spinning around its polar axis and the angular velocity of this spin increases as the core gains mass and compacts. It is possible that the core of our Sun might have reached a critical angular velocity in the past leading into a catastrophic event or events that might have resulted in partial fragmentation of its core near its equator. These fragments might have spun off capturing and carrying away some gaseous materials with them forming the planets either simultaneously or consecutively. (The fact that both our Sun and the Earth are rich in hydrogen might be a clue in this regard) In turn, these planets might have formed their own satellites going through a similar process while their cores are still in a partial plasma state. The important criterion to remember here is the coincidence of the orbital planes of the planets with the equatorial plane of the Sun. As for each planet within the solar system, same criterion applies to the orbital planes of the satellites and the equatorial plane of the host planet. With that criterion in mind, the only Pluto and its satellite Charon may have been captured planetary objects in our solar system since Pluto's orbital plane is significantly different than the equatorial plane of the Sun, . It is possible that the rings of Saturn might have formed while the planet is still in plasma state and might have remained in its close orbit as their cooling continued.

Once separated from the Sun, planets go through a number of phases depending upon their composition. However, their evolutionary phases by no means may be similar since their physical and chemical compositions vary significantly. Furthermore, their varying distances from the Sun might have played an important role in their cooling rates affecting their current cosmic status. Furthermore, their initial exit velocity from the Sun's core might have played an important role in determination of their current position within our solar system.

4.4 Blue Planet

After separating from the Sun's core, our home planet probably continued to its spin while still in a partial plasma state obeying the gravity and radiation field of the Sun. Nothing more than a fire ball ferociously burning through the Universe, it probably settled in an orbit which might be somewhat different than its current orbit since at the time it might have had a somewhat different overall mass and energy levels. If this is the case, despite our current belief system, the orbits of the planets in the solar system may be continuously changing over a cosmic time scale. How assuring.

Depending upon its distance from the Sun and the Sun's volume at the time, our Earth begins a long and unsteady period of cooling and differentiation. While heavier atoms drift towards its gravity center, lighter atoms make up the surface and gaseous atoms rise above the surface. As the Earth's core gains more mass, it is spinning faster than the surface of the Earth, very much like the Sun. While it might have been several million degrees hot at the time of separation from the Sun, our Earth's temperature begins to drop, relatively quickly at first but slower as the cooling goes on. While at the time, our Earth might have been a Gamma Ray source initially, as the cooling takes hold, it is now only a few thousand degrees hot and looks very much like a fire ball with an intense radiation signature, probably in the X-Ray range. It's further cooling results in formation of islands of crust on the surface, floating freely on the surface as directed by the currents of hot lava boiling just below the surface. This is the beginning of formation of continental plates, although they might have been entirely different than what they are today.

As the solidification of the Earth's crust takes hold, segments of the crust are now much thicker, each probably being several hundred feet thick, after having joined other segments of the crust floating near-by and covers the entire surface of the Earth. Thickening of the crust adds weight to it making it apply greater pressure on the lava below while trying to sink deeper. This pressure increase pushes lava through the cracks and joints of the crust, allowing the formation of volcanoes, helping accelerate the cooling process of the magma under the crust. As crust becomes thicker, may be several miles thick, surface is now much cooler, however, radiation field created by the core and hot lava below the surface accelerates atoms that make up the gaseous atmosphere above, creating plasma winds of charged gas atoms while creating a light displays of constant lightning and radiation storms. As volcanic activity continues at a high rate, more subsurface material transported to the surface from below, ever increasing the weight of the crust, forcing it to sink further. As the sinking accelerates, segments of the crust are laterally pushed against each other, creating anticlines, faulting and formation of mountains. At this stage, our planet has no longer a fairly flat surface, but has young mountain ranges as well as horsts, grabens, overthrust zones, deep ravines, canyons and vast depressions. Figs 4-11 thru 4-15. While Sun's radiation continues to modify the atomic composition of its gaseous atmosphere, its core, under immense pressure and heat, forms the magnetosphere to shield it from the Sun's radiation and solar flares. Once the surface is sufficiently cooled to allow accumulations of water and formation of seas, growth of prehistoric plants and animals takes place and the forces of geologic processes are established. Our blue planet has come a long way since its fiery beginnings and hopefully it has a long way to go. We can help by planting trees, preserving its eco systems, controlling man made pollution and eliminating deforestation.

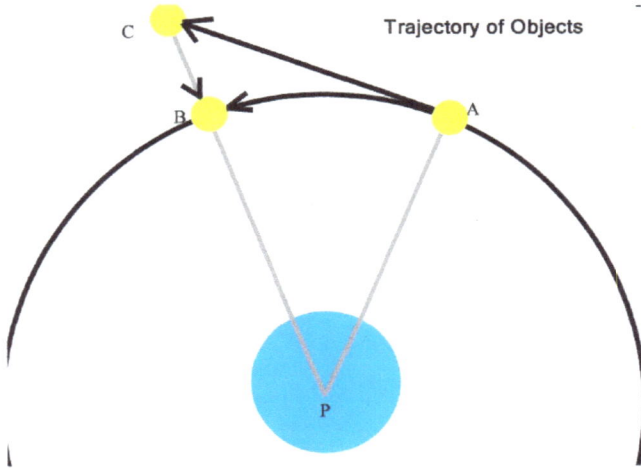

Trajectory of Objects

Fig: 4-1. The object at point A will arrive at point C instead of point B without the gravity pull of object at point P. When PA is equal to PB, the object at point A is said to be in orbit around object at point P. The object at point P is assumed to be stationary. Illustration by Stephanie C. Houston.

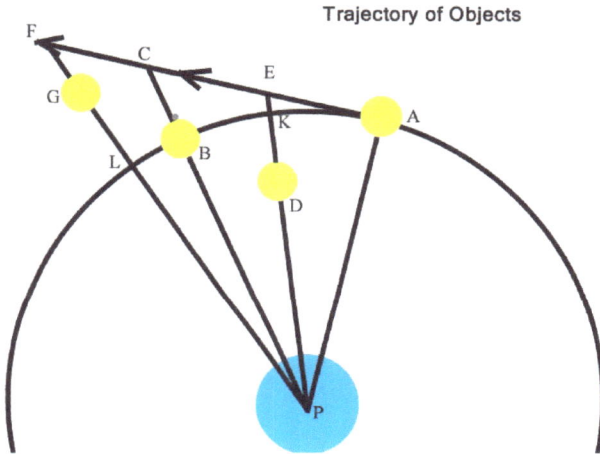

Trajectory of Objects

Fig: 4-2. The object at point A arrives at point B within a given time unit if the gravity pull of object at point P is equal to centrifugal force created by the orbital velocity of the object at point A. Since AP is equal to BP, the object at point A is in orbit around object at point P. The same object with slower orbital speed will arrive at point D instead of point K since the centrifugal force is smaller than the gravity pull of the object at point P. In this case, the object is said to be falling towards object at point P. On the other hand, the object at point A arrives at point G when its orbital speed is faster than object at point B since the gravity pull of object at point P is smaller than the centrifugal force. In this case, the object at point G is said to be reaching a higher orbit. The object at point B is assumed to be stationary. . Illustration by Stephanie C. Houston.

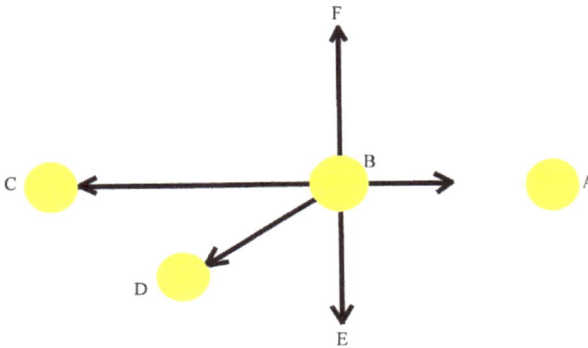

Movement of Objects

Fig: 4-3. The object at point B is being slowed down by the object at point A. With its slower speed, the object at point B arrives at point D instead of point C. BE indicates the direction of the central mass. . Illustration by Stephanie C. Houston.

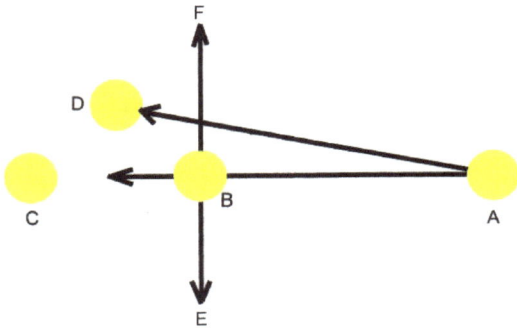

Movement of Objects

Fig: 4-4. The object at point A is being accelerated by the gravity pull of the object at point C so instead of arriving at point B, it reached up to point D. BE indicates the direction of the central mass. . Illustration by Stephanie C. Houston.

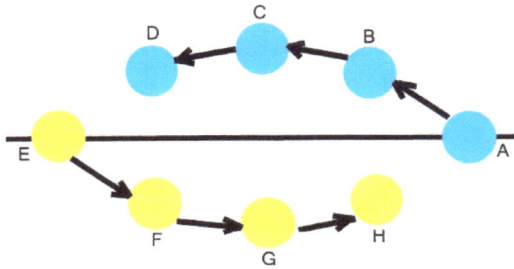

Braided trajectories of two similar sized
spherical objects with equal masses.

Fig: 4-5. In this figure, the object at point A is being accelerated by the gravity pull of the object at point E. It gains speed and reaches to point B above the shared mean orbit for the two. Similarly, the object at point E is being decelerated by the gravity pull of object at point A. It loses speed and falls to point F below the shared mean orbit. This interaction continues until object at point A reaches point C and the object at point E reaches point G. At this point the forward acceleration of object at point C and the deceleration of object at point E ends since the object at point C is pulling the object at G upward preventing it from falling further while itself stops climbing higher. From this point on, object at point A is being decelerated and pulled down while object at point E is being accelerated and pulled up. At the end, the object at point A reaches point E while object at point E reached point A only to repeat this cycle over and over again. The result of this interaction is a braided trajectory for both objects. . Illustration by Stephanie C. Houston.

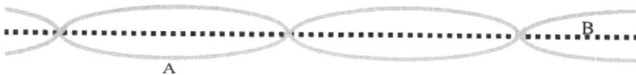

Braided trajectories of the objects
In this Diagram the curvature of the orbit is omitted.

Fig; 4-6. Braided orbital trajectories shown here repeat one after another. Illustration by Stephanie C. Houston.

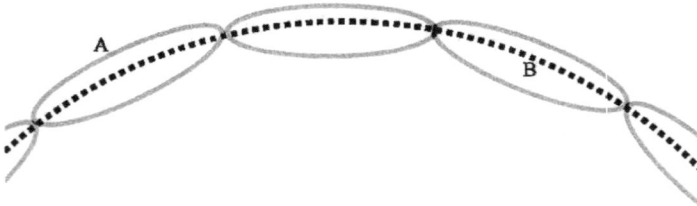

Braided trajectories of the objects

Fig: 4-7. In this illustration, the curvature of the shared orbit is exaggerated. . Illustration by Stephanie C. Houston.

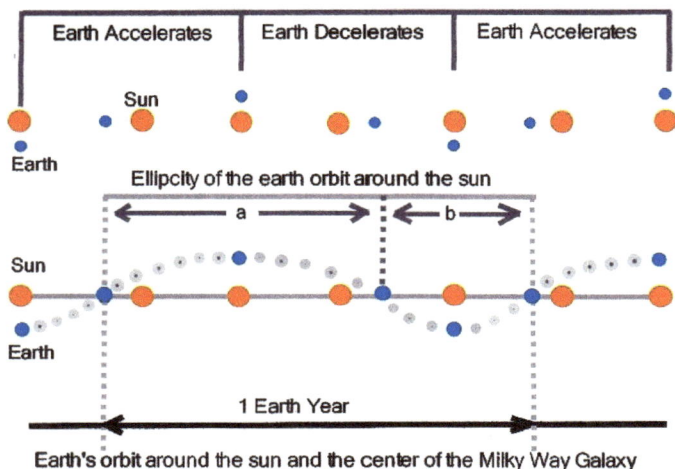

Fig: 4-8. Sun's orbit around the center of Milky Way Galaxy is presented as a straight line for all practical purposes because of its immense distance in between. The ratio of a and b represents the ellipticity of Earth's orbit around the Sun. However, we must remember that Earth actually is in orbit around the center of the Milky Way Galaxy and the Sun is not stationary. This presentation also applies to all other planets within the solar system and beyond as well as satellites of the planets such as Moon around the Earth. Based on that, what do you think the orbit of Halley's Comet will be like around the center of Milky Way Galaxy? Illustration by Stephanie C. Houston.

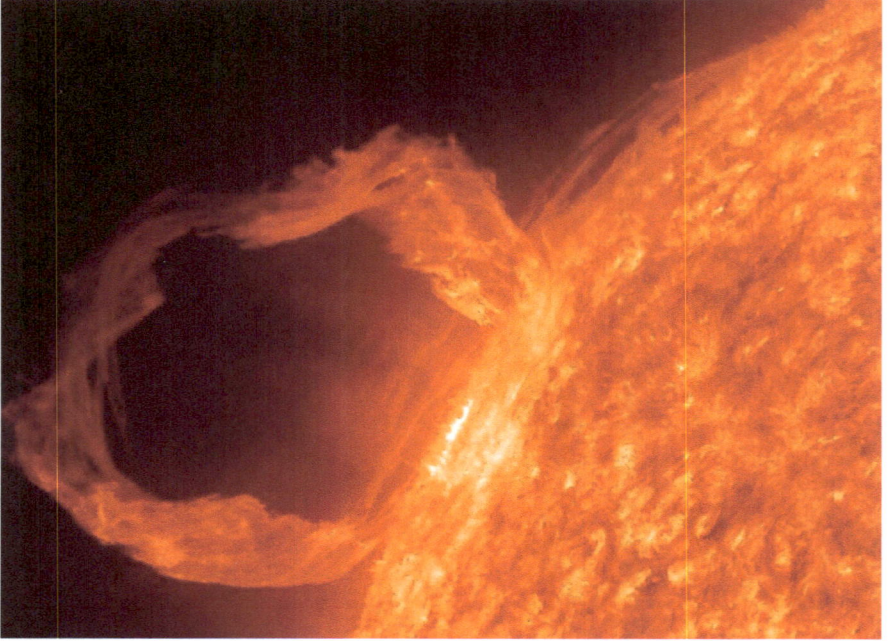

Fig: 4-9a. Solar eruption. Highly charged atoms with no mass shot up into space since Sun's gravity has no effect on atoms with no mass. As the eruption rises above Sun's surface, atoms begin to cool off developing mass. Once enough mass is formed, heavier atoms are pulled back into the Sun to repeat the cycle over and over again. Picture clearly shows that the temperature distribution at the Sun's surface in not even and the Solar eruption seems to be associated with a brighter region, presumably associated with higher temperatures. Image courtesy of NASA.

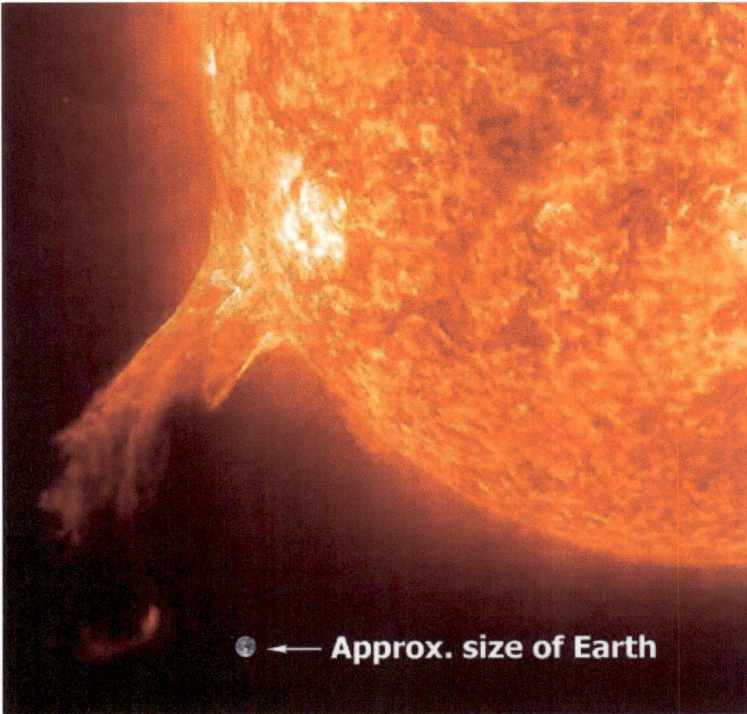

Fig:4-9b. Another example of solar eruption. These type of eruptions are
sometimes referred as a Solar Prominences. Image courtesy of NASA.

Fig: 4-9c. A solar flare. Image courtesy of NASA.

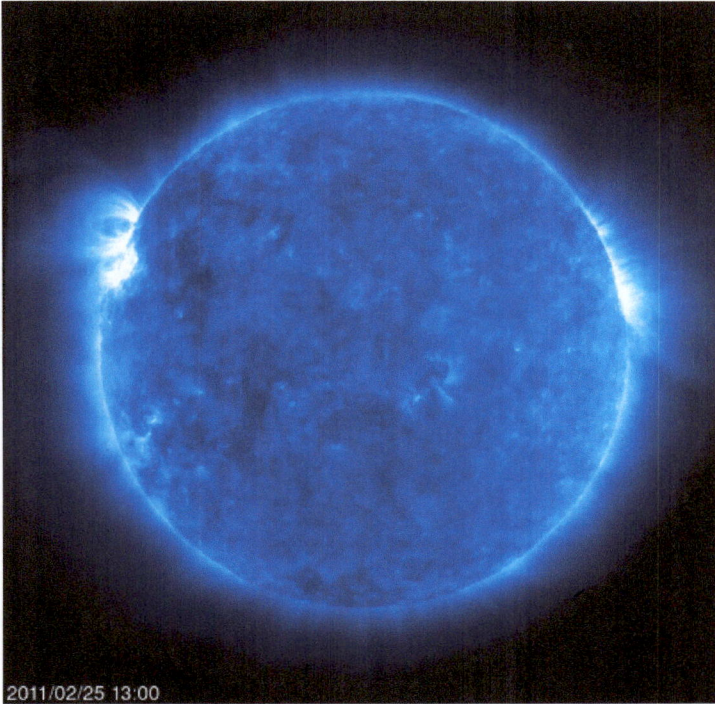

Fig: 4-10a. The sun at a closer distance will appear blue. Original image courtesy of NASA. Digital processing by Stephanie C. Houston.

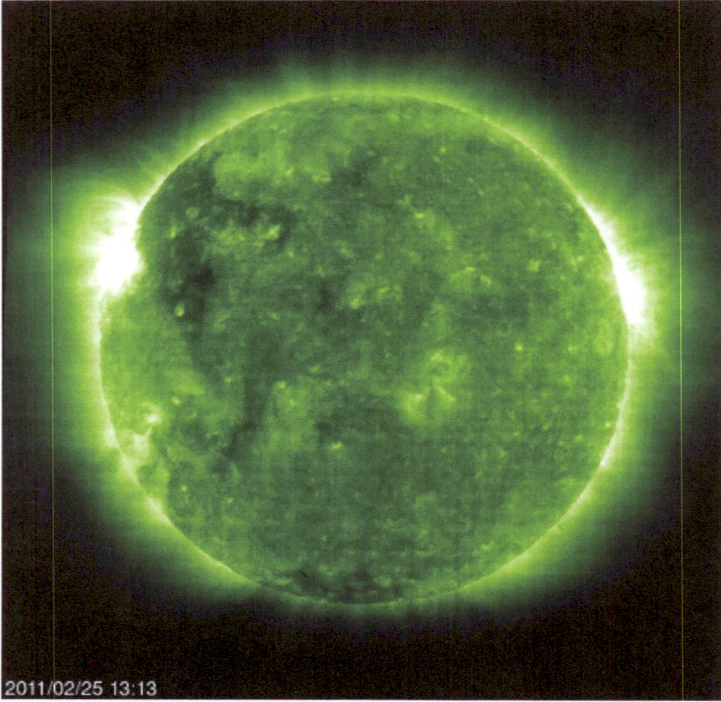

Fig: 4-10b. As the distance from the Sun increases, the Sun appears green. Original image courtesy of NASA. Digital processing by Stephanie C. Houston.

Fig: 4-10c. Further away, the Sun appears yellow. Original image courtesy of NASA. Digital processing by Stephanie C. Houston.

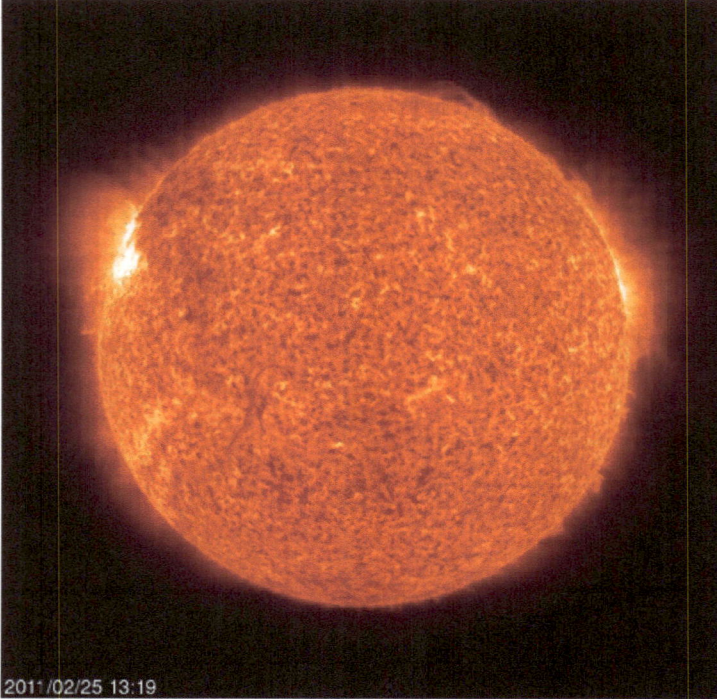

Fig: 4-10d. Moving further away, the Sun first appears red, than as a source of infrared, microwave and radio waves as a function of increasing distance. One correction is needed with these images. As we move away from the Sun, not only its color changes, but it also becomes smaller and smaller to the naked eye, a feature is not reflected in above 4 images. The effect of distance is approximated in the following image. One point of interest; image clearly shows that the temperature distribution on the Sun's surface in not uniform. Image courtesy of NASA.

Fig: 4-10e. Approximated effects of distance on Sun's appearance. Not to the scale. Photo montage by Stephanie C. Houston.

Fig: 4-11a. Cooling lava from a recent volcanic eruption forms a crust. A similar process may be responsible for the formation of Earth's initial crust at the very beginning. Source; Wikimedia Commons. Copied under Free Content and Public Domain License..

Fig: 4-11b. Ancient lava. This example is from New Mexico. Volcanic eruptions allow material below the Earth's crust to rise to the surface. After cooling and going through a weathering process, volcanic lava becomes a mineral rich fertile soil. Notice presence of plants. Photo courtesy of the author.

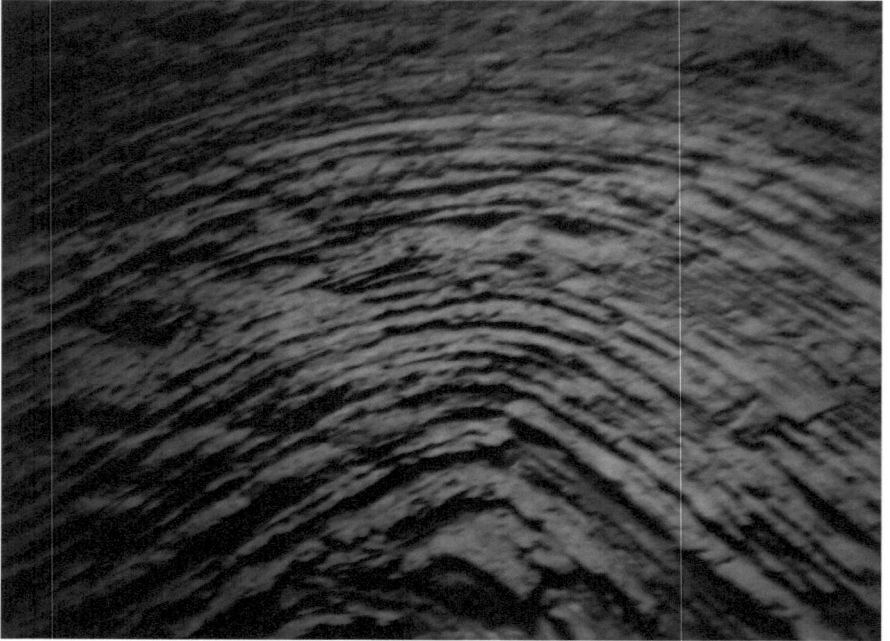

Fig: 4-12. An example of an anticline. Anticlines are usually formed when lateral forces compress the sedimentary rocks from both ends lifting up the central portion of the rock formation. This example is from Wyoming. Photo courtesy of the author.

Fig: 4-13. Normal Faulting. This particular fault image is from Arizona. Normal faulting happens when one side of the fault plane slips down when its support of underlying rock formation is no longer strong enough to carry its weight. Photo courtesy of the author.

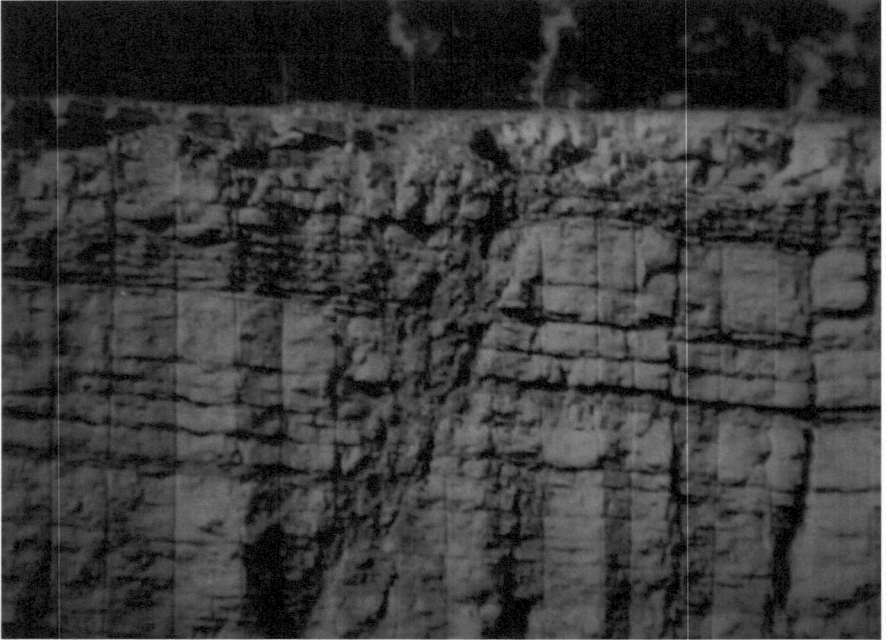

Fig: 4-14. Another example of normal faulting. This example is from West Texas. In this particular image, left side have slipped down. Fault plane is a zone of deformation and can be clearly seen. Photo courtesy of the author.

Fig: 4-15. An example of Graben. This particular example is from West Texas. Central portion of the rock layers has collapsed while both sides appear elevated. Photo courtesy of the author.

Fig: 4-16. An overthrust zone. This example is from Pennsylvania. In overthrust zones, extreme lateral forces push rock layers on top of each other. Photo Courtesy of the author.

Fig: 4-17. A Gas Giant in some distant galaxy, millions of light years away from our blue planet? Perhaps, but this one is an example of a sand stone, cut into a sphere, from painted desert in Arizona. Darker lines are due to presence of iron oxide. Photo Courtesy of the author.

CHAPTER 5

NEBULAE

5.1   Factories of atoms

Let us for a moment suppose that we do have a massive positive electric charge floating somewhere in the great vastness of the Universe and is attracted to a similarly large negative electric charge near by. As these two charges drift towards each other, they eventually contact and atoms form very much like oil droplets in water along the contact zone. Fig: 5-1a, b, c, d, e & f. As the atoms form, the process of atomic creation slows since the newly formed atoms form a barrier to further creation of new atoms. We call these massive electric charges, "Nebulae" as opposed to their being currently referred as gas clouds and dust in some reference books. Nebulae are visible to telescopic observations, because newly formed atoms within have a radiation signature. If there are no atoms, there are no frequency and amplitude and therefore no radiation is observed. The size of the atoms that form within a Nebula varies, presumably from helium to uranium in addition to those we have not yet discovered. The transition from atoms of one element to the next is not discrete but rather transitional and as a result we can have an infinite number of isotopes between atoms of two adjacent elements in the Periodic Table.

A Nebula is a factory of atoms within the Universe. As more atoms form within a Nebula, it becomes more visible. As the time goes by in the cosmic scale, these atoms within a Nebula continue to move in different directions within a three dimensional space of electric charges since the potential field of electric charges is stronger near the contact points and weaker as the distance increases. This allows atoms to disperse from the contact zones where they were once formed. As their cooling begins, some of these atoms now have greater masses and as a result they have smaller repulsive forces therefore they are attracted to one another to form young stars. This is why, some scientific literature call these Nebulae as star factories. However, we will continue to call Nebulae as places where atoms are first formed.

As its electric charges are depleted by forming atoms and anti-matter, a nebula becomes mostly matter and very little electric charge. Its temperature is cooler now and consequently it no longer shines brightly. This final stage of nebula is characterized by the presence of large number of brightly shining stars. Based on this reasoning, it is possible to assert that a very large nebula might have been behind the creation of star cluster Pleiades, NGC 1432.

5.2 Antimatter

The atomic model we have presented in this book so far has its positive electric charges within and its negative electric charge outside of its shell. If our atomic model had its negative electric charge within and its positive electric charge outside of its shell, it will be its own antimatter. The decision of which electric charge goes within largely depends on the sizes of nebulae involved in the atomic creation. This line of thinking makes it possible to assume that there may be a segment of the Universe, in which, atoms have their negative charges within and therefore are called antimatter. The question then becomes what happens when atoms of matter and antimatter collide. In all likelihood, their atomic structures disintegrate, their electric charges are freed to form a new set of atoms and that allows Universe to avoid a premature destruction. However, energy is indestructible and that is in line with the Principle of the Conservation of Energy.

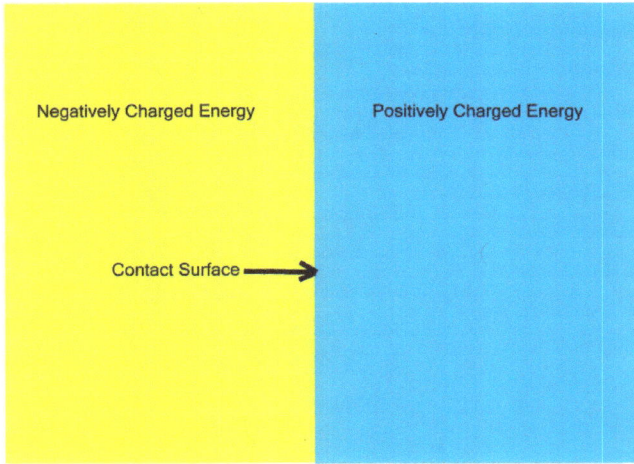

Fig: 5-1a. In this illustration, negative and positive energy are shown in their initial contact. However, the contact surface is probably is not a smooth one as depicted here for illustration purposes. Please note that this is a 2D presentation of a 3D phenomenon. Illustration by Stephanie C. Houston.

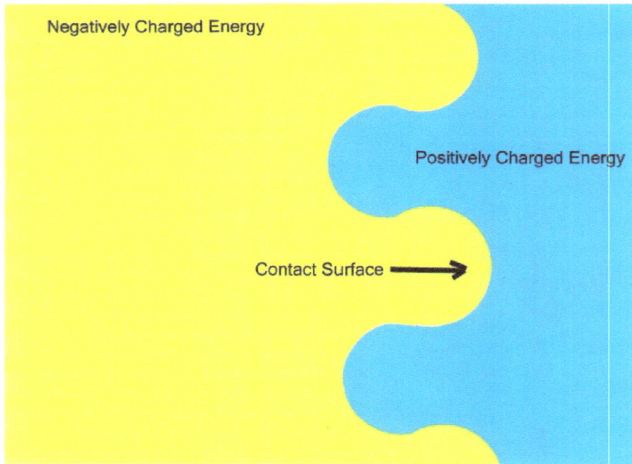

Fig; 5-1b. In this illustration, negative and positive energy are shown in the process of penetration into each other as a result of their attractive forces. Once again, the process is probably not a smooth one as depicted here. The purpose of the illustration is to simplify the process of forming of atoms. Once again, this is a 2D presentation of a 3D phenomenon. Illustration by Stephanie C. Houston.

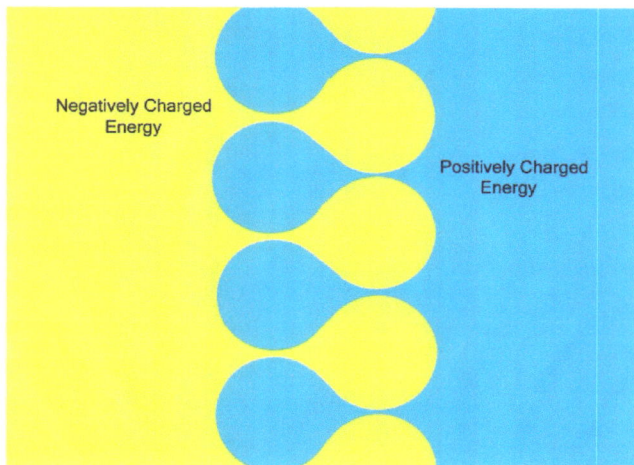

Fig: 5-1c. As the penetration of opposite charges advance, each electric charge takes form of a droplet engulfed by the opposite charge. Once again, this is a 2D presentation of a 3D event. Illustration by Stephanie C. Houston.

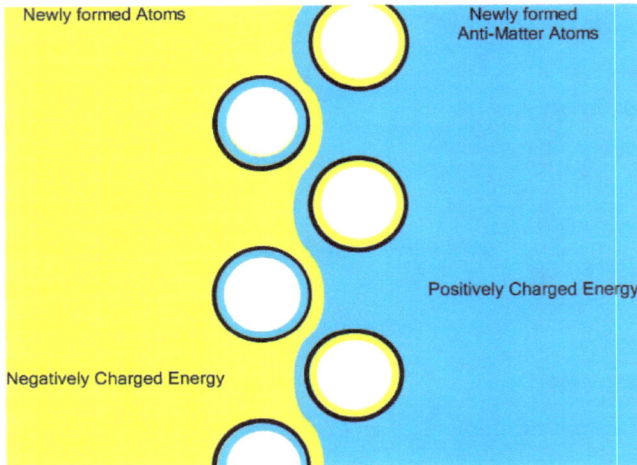

Fig: 5-1d. In this illustration, first set of atoms are shown formed with their spherical shape complete and their mass shown in black and in initial stages of development. These atoms will initially float in a sea of electric charge that clings to their shell. Once again, please try to visualize this 2D presentation in 3D. Illustration by Stephanie C. Houston.

Fig: 5-1e. After the completion of first set of atoms, contact surface continues the process to create the next set of atoms, again in 3D. Illustration by Stephanie C. Houston.

Fig: 5-1f. As more atoms are formed, billions and billions of atoms occupy the space near the contact surface forming a brightly shining space near the contact surface. These newly formed atoms slowly drift off the contact surface since the intensity of positive and negative electric charges are at their highest near the contact surface and this uneven energy field in 3D provides a force necessary for the drift. These atoms will gain more mass as they cool off and in doing so they will form millions of stars floating in the remnants of the nebula they are formed from. Again, please remember that this is a 2D presentation of a 3D event. Illustration by Stephanie C. Houston.

Fig: 5-2. NGC1976 Orion Nebula. M42. Orion Nebula is a young nebula with massive amounts of energy available for the formation of atoms and stars. It is brightest at the contact points of negative and positive energy. Color variations within are probably the results of different energy levels of its contents. Image courtesy of NASA/STScI.

Fig: 5-3. NGC 1952 Crab Nebula. M1. Crab Nebula is said to be a remnant of a supernova explosion witnessed by Chinese and Arab astronomers in 1054, expanding at about 1500 kilometers per second. However, dendritic structures crisscrossing the center of nebula contradict that assessment since an explosion would have been more consistent with an empty core. It is possible that the dendritic structures are made of newly formed atoms and plasma engulfed in a high temperature cloud of energy that may be propelling the nebula to expand. Image courtesy of NASA/STScI.

Fig: 5-4. NGC2074 is the Large Magellanic Cloud and is an example of a more mature nebula. It has large numbers of stars already formed within and exhibits large variations in temperature. Darker material in the foreground suggest that temperatures at its fringes too low to generate radiant energy and are probably grey or dark matter illuminated by the high energy regions behind. Image courtesy of NASA/STScI.

Fig: 5-5. Supergiant Star V383 is described as "Light Echo" by astronomers. It is probably a Nebula in its final stages after creating a large number of stars during its early stages as evidenced by the presence of large number of stars in its near vicinity. The relatively darker presence around its core suggests existence of cooler temperatures and probable an end to atom and star forming processes. Image courtesy of NASA/STScI.

Fig: 5-6. Tarantula Nebula is a middle aged nebula full of newly formed atoms and stars. Image courtesy of NASA/STScI.

Fig: 5-7. Star Cluster NGC290 was probably a large vibrant nebula initially, millions or billions of light years across. Image courtesy of NASA/STScI.

Fig: 5-8. Carina Nebula is a young nebula in the early stages forming of atoms. Bright areas of the nebula are the contact points of negative and positive energy. Image courtesy of NASA/STScI.

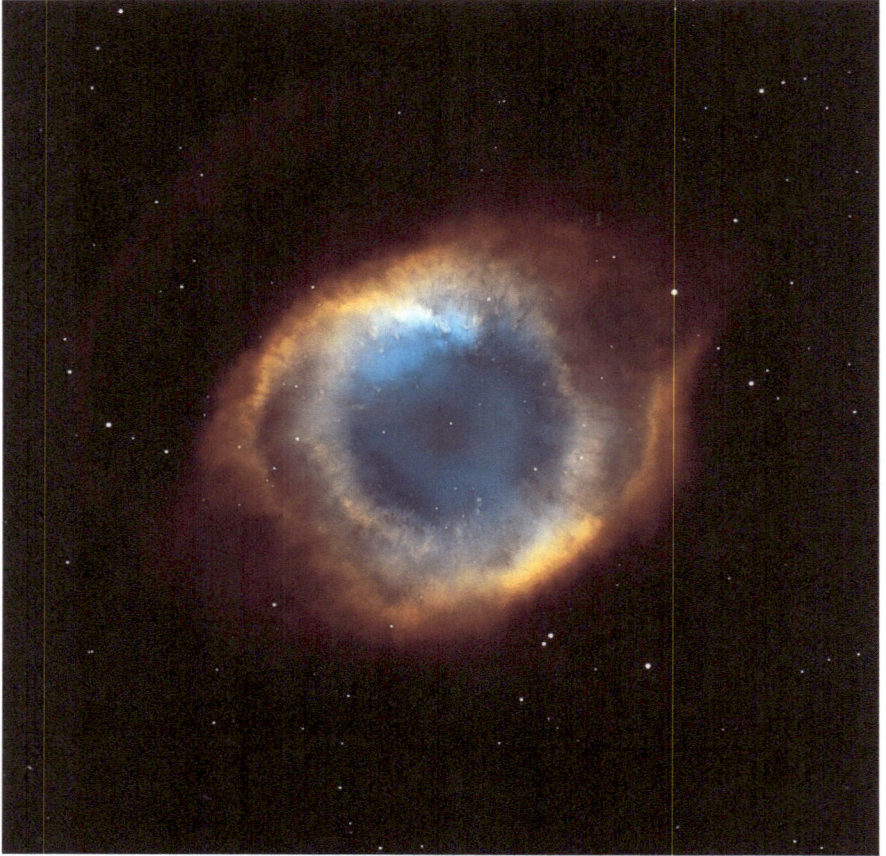

Fig: 5-9. Helix Nebula is probably a shock front of a supernova explosion. Supernova explosions are part of the cosmic cycle that transforms matter into energy. Please note the lack of large number of newly formed young stars in its near vicinity. Image courtesy of NASA/STScI.

Fig: 5-10. Mystic Mountain is a gathering of positive and negative energy shadowed by a presence of cooler gray matter at the bottom of the image. Image courtesy of NASA/STScI.

Fig: 5-11. 30 Doradus is a majestic nebula shown here in a composition of ultraviolet, visible and infrared light. Its star forming regions, clearly visible on the right hand side of the image, is home to millions of young stars shining brightly with a blue tint. Image courtesy of NASA/STScI.

Previous page;

Fig: 5-12. NGC 6302 Bug Nebula. Contact point of positive and negative energy is clearly pronounced by the intensity of contact and its bright radiant energy. Image courtesy of NASA/STScI.

CHAPTER 6

COSMIC CYCLE

6.1 Red Shifted Blues

In 1912, while working at Lowell Observatory in Flagstaff, Arizona, Astronomer Vesto Marvin Slipher (1875-1969) made a profound observation. He noticed that the absorption lines for a particular element within the spectrum of light received from distant Galaxies shifted towards the red end of the spectrum therefore appearing to have been imbedded in a lower frequency than it would otherwise have. He called this phenomenon "Red Shift". After arriving at Mount Wilson Observatory near Pasadena, California in 1919, Astronomer Edwin Powell Hubble (1889-1953) studied the work of Slipher and concluded that the red shift is a result of Doppler Effect in propagation of electromagnetic radiation similar to that of acoustic waves travelling in the air when the source of acoustic waves is moving away from an observer. As a result, he further asserted that the Galaxies must be moving away from our position of observation, faster if the distance is greater. This in turn, provided the basis for the current Theory of Big Bang, initially proposed by Georges Henri Eduard Lemaître (1894-1966) in 1927 which stipulated that the Universe is expanding in all directions and the speed of this expansion is greater if the Galaxies are farther away. Furthermore, this expansion must have a beginning at which point the Universe was condensed in a finite small super dense point, a point smaller than the point of a needle.

6.2 Seismic Acoustic Waves

An earthquake is a source of seismic energy which initiates propagation of seismic acoustic waves in all directions within our planet since the Earth acts like an elastic medium for this propagation. However, seismograms, the records of these seismic waves at observation points at different distances from a given seismic source, clearly show that these waves are subject to attenuation which is, by definition, a decrease in their energy levels, represented by their frequency and amplitude spectra, primarily as a function of the distance travelled. Attenuation is valid for all types of energy propagation including electromagnetic radiation since it is primarily an inverse function of distance travelled and can be expressed by $1/r^2$ where r donates the distance. It is clear from this line of reasoning that attenuation is also behind the red shift observed by Slipher within the spectrum of light received from distance Galaxies and the greater the distance, greater the red shift. At this point it is possible to assert that red shift is just a manifestation of attenuation of electromagnetic radiation by distance. If this line of reasoning of ours is correct, then Galaxies in the Universe are not flying apart, plunging away from our point of observation in an ever expanding Universe and gaining speed in doing so. The question then becomes what kind of force that exists in the Universe which propels this acceleration of expansion if indeed the Universe is expanding at such an increasingly alarming acceleration?

The amount of this presumed energy, often called Dark Energy, needed for the Universe to accelerate as it expands had to be either contained in an infinitely small dense point which gave birth to Big Bang or existed in the Universe prior to the Big Bang. If the energy powering the acceleration was contained in the infinitely small point which gave birth to the Big Bang, then energy decreases as an inverse function of the distance covered by the expansion therefore it can't sustain an ever increasing acceleration. If the energy powering the expansion of the Universe was already existed before the Big Bang happened, then the Big Band was not the beginning of the Universe.

There have been attempts to explain the existence of such a cosmic force but without any reasonable success. Furthermore, if the Universe has started as a very dense finite small point, what existed in the Universe before that which resulted in this dense small point? This line of reasoning leaves me no choice but to conclude that the Big Bang Theory is not fully proven to be valid within the current realm of our scientific knowledge.

6.3 Cosmic Cycle

If the Universe is not expanding as stipulated by the Big Bang Theory, do we have a static Universe as has been proposed in the past or are there other plausible explanations? We know that there are Black Holes in the Universe and they gather mass to become either Galaxies or Supernovas. These cosmic bodies are stages where mass turns into energy either by catastrophic events or cosmic process. This newly generated energy then forms galaxies, nebulae, stars and other cosmic wonders until they deplete their energy and become new formations of mass, which in turn become black holes starting a new cycle in the cosmic evolution. It is clear from this line of reasoning that our Universe is constantly changing between a state of energy and a state of mass and this process is an inherent characteristic of our universe. Based on this, it is plausible to assert that our Universe is both expanding and contracting depending on the nature and location of cosmic activity it is subjected to. But, if that is the case how do we find out how old is our Universe?

6.4 The Age of Universe

Scientists have determined that the Universe is approximately 13.7 billion years old by obtaining precise measurements of the Cosmic Microwave Background (CMB) radiation which has been estimated to be about 3 degrees above the absolute zero with minor fluctuations in different directions. However, distribution of mass and energy in Universe is hardly uniform. Efforts to map Universe show that the known structure of the Universe is highly random in nature and that is a challenge.

On the other hand, if we could have a firm grip on the size of the Universe, we could have that distance divided by, say, the speed of light, assuming that the Universe was expanding at that constant speed. If that is the case, then we could have reached a number expressing the age of Universe. But there are problems. How do we know that the Universe expanded at the speed of light? What if it is expanding with an ever increasing acceleration? Hubble calculated the speed of receding galaxies by using a ratio of red shifts and called those apparent velocities. Are apparent velocities calculated from different red shifts good enough to be used in our calculations if they are significantly different at various distances from our observation point?

Here we will try another approach to estimate the age of Universe since in this book I have disagreed with the Theory of Big Bang. Going back to our Cosmic Cycle, if the Universe has started as a single or multiple pairs of pure energy in form of a positive and negative energy charges in the cosmos, then our Universe was 100% energy and 0% mass at birth. We could say that our Universe is 0% old. If we can estimate the amount of mass and energy that exist in the Universe today and calculate a ratio, then we can say our Universe is say, 80% old, meaning that in our Universe, we have 80% matter and 20% energy. Of course, we must remember to factor in Einstein's (Albert Einstein; 1879-1955) constant, which is the ratio of conversion between energy and mass expressed as c^2 in $E=mc^2$, which is the second power of speed of light. So will our Universe be considered dead if it is 100% mass and 0% energy? But this creates an interesting dilemma. We know that in our Universe, mass and energy is constantly converted between the two, mass to energy as well as energy to mass, and the ratio in between the two is constantly changing. Is it possible to say that our Universe can be 80% old today in some parts of it but only 50% old in some distance future? Is the age of Universe uniform at every point in Cosmos? What if it is not? How that can change our perceptions, if one day we can have the means to travel from a part of Universe that is 90% old to a part that is only 50% old? I will leave these questions for you to ponder.

6.5 Conclusions

The picture that emerges from our line of reasoning is different than the generally accepted principles of our current scientific realm. What we have proposed here is a Universe that started out as a sea of 100% energy in a three dimensional space and is continuously changing between a state of mass and a state of energy. Will it end when it becomes 100% dark matter although cosmic cycle we have proposed contradicts that line of reasoning?

In this book, we have for the first time ever proposed a dynamic atomic model that can be manufactured in the Universe with great ease and can be the building block of all forms of mass, energy and life in Cosmos. We have suggested a Cosmic Cycle in which matter and energy are continuously in transition between the two and a Universe both expanding in some parts and contracting in some others. We have even attempted to find out the age of our Universe, which I believe is extremely important in understanding our past and future as Cosmic Travellers in the words of Carl Sagan. We asserted that life is all frequency and amplitude and has its own unique spectra which are unique to all life forms. We have challenged The Big Bang and tried to clarify our misperceptions about Red Shift as defined by Edwin Hubble. We also have for the first time tried to understand magnetism and offered a plausible suggestion for the existence of The Unified Field Theory. These are significant deviations from the current realm of scientific knowledge and they invite further investigations.

But we must pause here now to take a break since I believe we have talked enough about my ideas, my perceptions, my theories and my conjectures. It is your turn now.

Fog: 6-1. Absorption lines from two celestial objects. The object whose spectrum is at the bottom is said to be farther away from the one whose spectrum is at the top. Source; Wikimedia Commons. Copied under Free Content and Public Domain License.

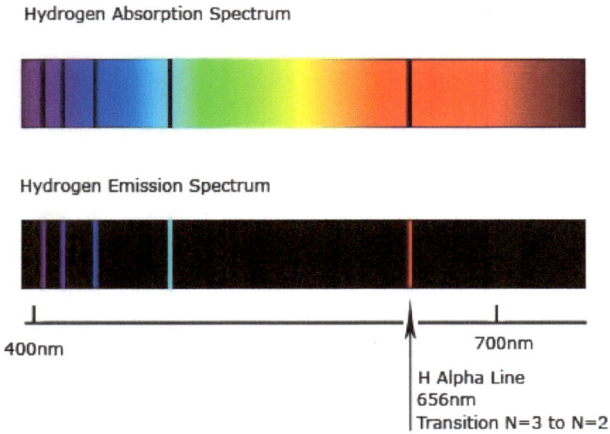

Hydrogen Absorption Spectrum

Hydrogen Emission Spectrum

400nm

700nm

H Alpha Line
656nm
Transition N=3 to N=2

Fig: 6-2. Absorption and emission lines of Hydrogen. Source; Wikimedia Commons. Copied under Free Content and Public Domain License.

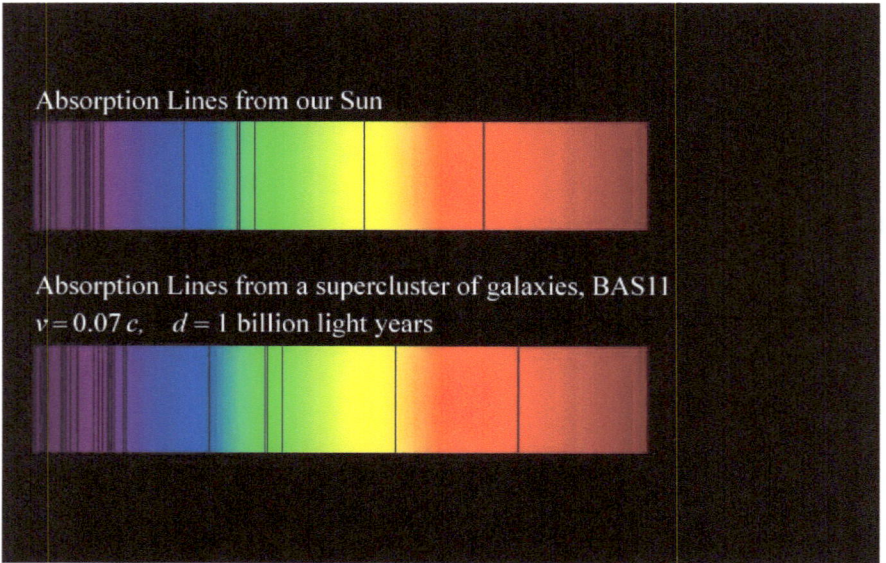

Fig: 6-3.Simulated comparison of the Sun's absorption lines with absorption lines from a distant supercluster of galaxies. Image courtesy of Prof. Dr. Harold T. Stokes, Brigham Young University.

APPENDIX 1

GRAVITY WAVES

Gravity is said to be a force of attraction between two masses of any size. This force is proportional to the product of the two masses and inversely proportional to the square of the distance between them as asserted by Sir Isaac Newton (1643 – 1727). It is the reason why objects have weight and why they do fall back to the Earth. It is also the reason why planets of our Solar System orbit our Sun. Without the force of gravity, the Universe will be a much different affair.

More precisely, however, gravity is a three dimensional potential field surrounding a mass of any size. Naturally, greater the mass, greater the magnitude of the potential field as expected. However, having a mass alone is not enough to create a gravity wave. In order to create a gravity wave, a mass must have motion.

Let's for a moment consider a large planet orbiting a massive stationary star with a significant gravity pull. Let us also assume that we are within a distance of observation, in cosmic scale, from the two. The gravity pull we will feel will be the summation of the gravity pulls of the planet and the star together as a binary system. Since the star is stationary in reference to our point of observation, its gravity pull will appear unchanged during our observation. However, as the planet in question moves around the star in its orbit, the combined gravity pull we will feel and measure will change as a function of the planet's motion. When the planet is directly in between us and the massive star, we will feel the greatest gravity pull of the binary system because the force of gravity of the planet is the greatest since it is at its closest point to us. When the planet moves to the other side of the star, it is at its farthest point from us therefore its gravity pull is the weakest. So the combined gravity pull of the two is now smaller. If we continuously plot our measurements of the combined gravity pull of this binary system at our station of observation as a function of the planet's move around the star over a period of time, we will end up plotting a gravity wave.

The Universe is full of gravity waves because it is full of billions of massive celestial objects moving around in a cosmic dance. Yet gravity waves are the most difficult to detect because of immense cosmic distances that can be best measured in light years. Since these great distances reduce the effects of gravity by the inverse square of distance, gravity waves quickly become infinitesimal in magnitude in the great vastness of the Universe which should explain why they are so difficult to detect.

But what happens if suddenly a large celestial object converts its mass into energy as in a supernova explosion? Anytime a celestial object with mass ceases to exist, its three dimensional gravity potential also ceases to exist since gravity is an attribute of mass and nothing else.

APPENDIX 2

OLBERS' PARADOX

One of the questions that preoccupied astronomers and the philosophers in the 16th century was about the night sky. They had wondered why the sky was black at night yet stars were bright and could be easily seen on a clear night. With so many bright stars dotted the sky above, why, they have enquired, they don't see a sky that is full of light once the Sun goes down. This was an apparent contradiction with their belief in an essentially infinite and ageless Universe. German astronomer Heinrich Wilhelm Olbers (1758–1840) was one of those who also considered this apparent paradox. Of course, present day astronomers believe that the Universe has an age, 13.5 billion years at the last count and perhaps subject to change in the future, as well as it might not be limitless at all. So some astronomers are hard at work trying to unearth the answers to this seemingly contradictory observation. One of their conclusions so far was based on the present day interpretation of red shifted absorption lines which implied that different amounts of red shift can cause the visible light to become invisible despite the fact that we have optically imaged galaxies that are almost 13 billion light years away.

Our observations tell us that our Universe is full of electromagnetic radiation. As a result we know that our planet is constantly bombarded with Gamma Rays, X-rays, Microwaves and Radio waves of cosmic origin as well as visible light. We also know that all energy propagation in space is subject to attenuation meaning Gamma Rays attenuate as a function of propagation to become X rays, X-Rays to become visible light, visible light to become Micro Waves and Micro Waves to become Radio Waves. Hence, we only receive radio waves from the most distant objects in the Universe. Yet our eyes and brain can only detect, process and image part of the electromagnetic spectrum which falls within the frequency range of visible light meaning that a very large volume of the Universe is not visible to our most sensitive optical observation tools. There is a Universe far beyond our planet that we can not see yet we know it is there since it can be observed with radio telescopes. If our eyes were able to detect and see the entire frequency range of the electromagnetic spectrum, our night sky will be brighter than we can possibly imagine and perhaps more brighter than we would ever want it to be.

APPENDIX 3

ORGANIC MOLECULES

For a very long time, scientists have grappled with a fundamental question without much success. How do organic molecules, building blocks of life, form from atoms of known elements which are considered to be inorganic by definition?

In Chapter 5, we have proposed a model of how atoms are formed at the contact points of negative and positive energy freely floating in space. We also identified nebulae as factories of atoms. But how did organic molecules form from these newly hatched atoms in order to create building blocks of life?

Contact points of negative and positive energy are zones of extreme heat and as a result, atoms created in these environments are mostly energy and very little mass at the early stages of their existence. Since energy outside the shell of these atoms are the source of their repulsive force according to our atomic model, these atoms are driven away from contact points of negative and positive energy as well as from each other towards zones of lower energy and finally into the intergalactic space. As the cooling of these atoms takes hold, our atoms are now less energy and more mass and as such they have lesser amount of repulsive forces while gaining more of the attractive force of microgravity. It is this weak force of microgravity that allows atoms of all kinds to combine in random processes to form initial simple molecules. These initial simple molecules continue to interact with each other through random iterative processes to eventually become more complex organic molecules and building blocks of life.

Similarly atoms present in our home planet after its fiery birth might have gone through a similar process of cooling to allow formation of Earth bound organic molecules. If this line of reasoning of ours is indeed plausible, our Universe must be brimming with life as new organic molecules are being formed every second in its great vastness.

But is there any supporting evidence for this line of reasoning that molecules form within certain temperature windows during a cooling period? We all know that certain chemical reactions happen at room temperature but some others require certain degree of heating. This means that temperature levels play a necessary role in all chemical reactions since atoms can combine to form molecules if their attractive forces are greater than their repulsive forces. Accordingly the balance between repulsive and attractive forces of our atomic model is crucial in all chemical reactions since temperature is the only factor that controls that critical balance between atoms of all elements.

EPILOGUE

A book is a permanent record of a thought process. As such, I am responsible for every idea presented in this book. If, sooner or later, these ideas get to be proven wrong, then I will be happy to stand corrected. In the meantime, thank you for joining me in this exciting journey to reach an unknown destination in the great vastness of the Universe.

PROFILE

The author has received a four year degree in Geophysics from the Science Faculty of University of Istanbul in February 1966 before starting a lifelong international career around the world working for Geophysical Service and International Oil Companies, specializing in application of seismic methods and digital technologies to hydrocarbon exploration. He has resided and worked in 6 different countries on 4 Continents during his long career developing global perspectives and a greater understanding of humanity. In 1986, he became a US citizen and changed his name from Güven Demirseren to Robert Houston during naturalization process. In two separate letters to American Geophysical Union and The Society of Exploration Geophysicists, he first proposed that the Inner Core of our Earth may be spinning faster than its Crust. In 1997, scientists documented his prediction to be correct. Encouraged by this development, he extended his work into Astrophysics and published the first edition of The New Universe (ISBN; 1-57733-057-9) in 1999 in which he presented for the first time his theories about the birth and evolution of galaxies, internal structures of black holes and atomic spectra. He is currently a resident of Texas and working as a consultant. He could be reached at guvenli1@yahoo.com for inquiries.

www.ingramcontent.com/pod-product-compliance
Lightning Source LLC
Chambersburg PA
CBHW041304210326
41598CB00005B/24